IN DEEP WATER

The Anatomy of a Disaster, the Fate of the Gulf,
and How to End Our Oil Addiction

IN DEEP WATER

The Anatomy of a Disaster, the Fate of the Gulf,
and How to End Our Oil Addiction

Peter Lehner
with
Bob Deans

OR Books, New York

Visit our website at www.orbooks.com

First printing 2010.

Library of Congress Cataloging in Publication Data:
A catalog record for this book is available from the Library of Congress

British Library Cataloging in Publication Data:
A catalog record for this book is available from the British Library

Typeset by Wordstop, Chennai, India

Printed by BookMobile, USA

The printed edition of this book comes on Forest Stewardship Council-certified, 30% recycled paper. The printer, BookMobile, is 100% wind-powered.

paperback ISBN: 978-1-935928-09-6
ebook ISBN: 978-1-935928-10-2

10 9 8 7 6 5 4 3 2 1

For Nadine, Eliza, and Marina

This book reflects my desire to prevent future catastrophes to the human and natural world.

TABLE OF CONTENTS

Timeline

April 20 • BP's Macondo well blows out beneath a mile of water in the Gulf of Mexico, spewing oil and natural gas over the drill rig Deepwater Horizon. An explosion and fire on the rig, owned by Transocean, kills eleven workers out of 126 aboard.

April 22 • After burning two days at sea, the Deepwater Horizon sinks, shearing off the riser pipe connecting the wellhead to the rig and unleashing a torrent of crude oil and natural gas into the ocean.

April 24 • Remotely Operated Vehicles (ROVs) inspect the capsized rig and find two oil leaks from the well pipe along the sea floor and a third from the riser. BP says up to forty-two thousand gallons a day could be leaking; later revises estimate to two hundred thousand gallons a day.

April 30 • Defense Secretary Robert Gates mobilizes Louisiana National Guard to help with oil cleanup and removal and to protect critical habitat.

May 1 • Homeland Security Secretary Janet Napolitano names U.S. Coast Guard Commandant Thad Allen the National Incident Commander for the administration's continued response.

May 2 • BP begins drilling first relief well; estimates it will take ninety days to complete.

May 4 • Pentagon approves mobilization of up to seventeen thousand five hundred National Guard troops to help various states with spill (six thousand in Louisiana and Mississippi, three thousand in Alabama, two thousand five hundred in Florida).

May 5 • BP says it has stopped flow of oil from one of three leaks; doesn't change overall flow rate.

May 8 • Large cofferdam cap fails due to an excess of methane hydrate crystals forming inside the dome.

May 16 • Riser insertion tool tube inserted into leaking riser; picks up estimated 84,000 gallons a day.

May 17 • BP begins drilling second relief well.

May 20 • EPA directs BP to identify a less toxic dispersant within 24 hours and to begin using it within seventy-two hours of identifying the alternative. BP responds that it can identify no less toxic dispersant.

May 22 • Obama creates bipartisan National Commission on the BP Deepwater Horizon Oil Spill and Offshore Drilling.

May 24 • More than sixty-five miles of Louisiana shoreline impacted by oil. Secretary of Commerce Gary Locke declares a fisheries disaster for commercial and recreational fisheries in the Gulf of Mexico; NOAA has closed nineteen percent of American Gulf waters to fishing. Eventually thirty-seven percent of those waters would be closed.

May 26 • BP initiates top kill.

May 27 • National Incident Command's Flow Rate Technical Group estimates oil spill rate at between five hundred four thousand gallons and eight hundred thousand gallons per day.

May 29 • Top kill attempt ends in failure.

May 31 • NOAA extends northern boundary of closed fishing area. Closed area: 61,854 square miles; nearly twenty-six percent of the Gulf federal waters.

June 4 • BP places containment cap over leak source.

June 15 • Obama addresses the nation from the Oval Office: "Already this oil spill is the worst environmental disaster America has ever faced."

June 16 • BP pledges to create $20 billion escrow fund in White House meeting with Obama.

July 15 • BP caps the spill, ending the release of significant amounts of oil into the Gulf, after an estimated two hundred million gallons of crude oil gushed out over three months.

August 4 • A NOAA team estimates that seventy-four percent of the oil has either been captured from the wellhead, burned off, skimmed, dissolved in the water, dispersed or evaporated. That leaves twenty-six percent, the government says, washed ashore, buried in beaches or marsh, or drifting through coastal waters or deep oceans in plumes or sheen.

August 5 • Gulf area residents voice skepticism over federal claims. Even if true, some point out, it meant that one hundred million gallons—roughly half—of the oil remains in the water or along the coast.

Foreword

by Wendy Schmidt

Americans everywhere have been dismayed by what the BP oil disaster has done to the Gulf of Mexico and to the people living in that unique region. Our inability to effectively address the blowout, and its catastrophic aftermath, has been frustrating.

What exactly went wrong on Deepwater Horizon? And why does it matter? This is the story told here. *In Deep Water* describes the largest environmental disaster in American history, and explains what we can do to change our self-destructive ways. The surface oil may have dissipated, but in mid-August, scientists from the Woods Hole Oceanographic Institution identified an undersea plume at a depth of three thousand six hundred feet extending for more than twenty miles southwest of the well. In places, it is more than a mile wide and six hundred feet thick, traveling about four miles a day. This subsurface oil continues to pose a threat to the food chain of marine life.

Drawing on three decades as one of the nation's leading environmental lawyers, Peter Lehner offers an authoritative account of this ongoing tragedy, and puts a human face on its consequences, calling upon the powerful eloquence of voices from the Gulf. At a time when commentators and politicians are piling on BP, Lehner leavens his criticism with self-reflection, stressing the role all Americans play in this debacle as the world's largest consumers of oil. Oil is a resource that is valuable, even precious, he writes. It's expensive to find, dangerous to produce, and harmful to burn. We need to ask if there might be ways to use oil more wisely, to use less overall, and to begin seriously investing in development of alternative fuel sources.

In a story that could easily descend into despair, Lehner finds hope in our democracy's ability to heal its wounds, learn from its mistakes and emerge ever stronger from the fire. Having surveyed the landscape damaged by oil, he shows the way to higher ground, toward a future where drilling is safer and we drill less, as we reduce our reliance on oil.

While we envision a clean energy future, we need to be realistic, as Lehner acknowledges. With nearly four thousand oil platforms operating in the Gulf, some of which serve as many as two-dozen undersea wells in water up to two miles deep, it isn't a question of whether we'll suffer another oil spill someday, but rather, when. How can we be prepared to address the destructive legacy of the energy production system that has fueled our world for nearly a century?

Part of the solution lies in the development of breakthrough technologies for containing oil from blowouts like this one, and for cleaning up the oil quickly to protect life on land and in the sea from harm. That's why I'm sponsoring a prize to the research team that demonstrates the ability to recover oil on the sea surface at the highest oil recovery rate and the highest recovery efficiency.

Read this book. Pass it on to a friend. It will change forever the way you think about oil, the Gulf of Mexico and the people who live there. It may also make clear the challenges and opportunities we will find as we develop renewable energy resources.

Wendy Schmidt is a trustee of the Natural Resources Defense Council, co-founder of the Schmidt Marine Science Research Institute and president of the Schmidt Family Foundation. For more information about the Schmidt Prize, go to www.iprizecleanoceans.org.

Introduction

The Deepwater Horizon disaster has put three urgent questions before the country. What happened? How did we arrive at this point? And: where must we go from here? The way we answer these questions will have a profound impact on the future of the Gulf of Mexico, its people and, indeed, Americans everywhere. The implications for our economy and national security, our environment, and our entire way of life can hardly be overstated.

This book is an attempt to address those questions, providing answers where they exist, clues where they do not and guidance as to where we might find them. As the questions are urgent, the book is meant to be timely, a first draft of sorts. The narrative that we tell is the story so far, with much remaining to discover and learn. We hope the various analyses underway, as well as the president's commission investigating the blowout (on which NRDC President Frances Beinecke serves), will bring us all to a more thorough understanding.

On the face of it, what happened is clear: the fourth-largest corporation in the world drilled a well beneath a mile of water in the Gulf of Mexico. Operating at the frontier of knowledge in conditions more challenging than even deep space, this well was inherently dangerous. There were a series of judgments, operations and technological shortcomings running from the planning and design of the well to the effort to seal it that set the stage for disaster. When the well blew out on the night of April 20, equipment designed to be the last line of defense against disaster failed to shut down the well.

Eleven men were killed that night and more than two hundred million gallons of crude oil gushed into the ocean during the three months it took for the oil company, formerly known as British Petroleum, to cap the well. More than six hundred miles of coastline were oiled and a slick the size of South Carolina covered the fertile Gulf. Ocean, coastal and wetlands habitat and birds, fish, marine mammals, sea turtles and other animals and plants were damaged or destroyed. Thirty-seven percent of American waters in the Gulf were closed to fishing; thousands of watermen and others were thrown out of work; and the future of a complex and vital region was cast into uncertainty.

Our country arrived at this point because our demand for oil has driven companies like BP into deeper and riskier Gulf waters at precisely the time the political consensus had broken down for the public safeguards we need to protect our safety, health and environment. Our watchdog agencies, in short, were defanged. And so, just as companies pushed the outer limits of their technological capability and operational expertise to drill for oil in water up to two miles deep, the agency responsible for ensuring the industry's safety was steadily becoming compromised, losing its ability to keep up.

That agency, the Minerals Management Service, an arm of the Department of the Interior, was addled by rules and authority that were years behind the fast-changing offshore oil industry. The agency had five dozen inspectors to keep tabs on four thousand offshore platforms, some of which serviced more than two dozen wells, across the Gulf of Mexico.

The close-knit culture of the region, and a near-incestuous relationship between the agency and the industry it was supposed to oversee, created an atmosphere in which oil company engineers sometimes penciled in their own responses to inspection forms and federal inspectors later traced over those replies in ink. Offshore oil companies treated inspectors to private hunting expeditions, skeet shoots and fishing trips.

Beyond such cultural proclivities, there were institutional mandates as well. President George W. Bush and his vice president Dick Cheney were both oil company executives before they came to the White House. Waging war in Iraq, home to the fourth-largest proven oil reserves in the world, those men also put in place energy policies aimed at boosting domestic production. In the Gulf, that meant speeding the permitting process for offshore wells like the one BP lost control of in April, and waiving requirements for adequate environmental review and sufficient oversight of blowout response plans. And what was happening in the Gulf was reinforced by similar developments around the country. Most efforts to conserve or use energy more efficiently were stymied, in part due to the opposition of the oil, gas and coal companies. At the same time, agencies Congress had ordered to protect the public were underfunded, staffed with industry allies and intimidated by attacks in the press.

When BP's Macondo well blew out, the company had no idea how to stop the runaway well and no equipment in place to cap it. From the president on down, Americans watched in helpless fury as two million gallons of crude oil a day gushed into some of the most diverse and productive fisheries in the world. The Gulf is home to scores of species of fish, from the majestic bluefin tuna to the common carp, numerous endangered species, such as whales and sea turtles, and birds both migratory and resident. It is the source of seventy percent of the oysters and shrimp produced in this country and hundreds of millions of pounds each year of snapper, grouper, tuna and other seafood. Drilling for oil in the Gulf of Mexico, it turned out, is an activity that puts an irreplaceable resource at risk. There are only two rational responses: reduce the risk and reduce the need for the activity.

We can do both; we must do both—or else further catastrophe awaits. In this book, we show how we can make offshore drilling safer by investing in the safeguards we need, the institutions required to enforce those safeguards and the professionals we can count on to protect our safety, health and environment. I am not a technical expert and this book does not purport to offer engineering solutions: rather, the book suggests structural changes that will ensure that we have adequate technical experts with the ability and authority to protect us. And we show how we can cut our consumption of oil in half while reducing the threats to habitat and health.

The conclusions I draw are informed by a thirty-year environmental law career that began on the Gulf Coast. Working for the Sierra Club Legal Defense Fund, I helped bring a lawsuit against the federal government regarding the environmental impact of a planned bridge at Dauphin Island, in Alabama's Pelican Bay. We lost the case and the bridge was built, but I developed an abiding interest in this rich and robust region and the endless complexity of its wildlife and habitat. Working on the case also inspired me to go to Columbia University Law School, where I learned that, if we're right on the law and the science, we can accomplish an enormous amount for our environment. I took that adage to heart, creating the environmental prosecution unit at the New York City Law Department. Our first big case addressed New York's biggest oil spill, when an Exxon pipeline burst. The case resulted in major operational changes and wetlands restoration. I later headed the New York Attorney General's Environmental Protection Bureau, with an excellent staff of forty lawyers and ten scientists, where I saw both a wider range of fossil fuel energy impacts and opportunities for a cleaner future.

As executive director of the NRDC, I oversee the work of more than four hundred dedicated environmental advocates, lawyers and scientists supported by 1.3 million members and activists nationwide. From the early days of the BP blowout, we recognized the potential threat it posed to deep ocean, coastal waters, shoreline, wetlands, estuaries and bays, as well as to our air, inland fresh waters marine life, aquatic life and birds. This book pulls together the work of an entire staff of dedicated professionals, including dozens who have been directly involved in our oil spill response efforts from the opening days of the disaster.

The *Exxon Valdez* oil spill in 1989 focused the NRDC's attention on the risks of producing and transporting oil in the frigid Arctic waters far north of the Gulf of Mexico. Our advocacy succeeded in putting in place a twenty-year moratorium on drilling in the Arctic and many of our coastal waters, but we did not succeed in protecting the Gulf, which has seen decades of environmental degradation in the name of drilling and other commercial activities. Hurricane Katrina reawakened the nation to the decades of harm experienced in the lower reaches of the Mississippi delta, and the impact that deterioration had been having on the Gulf. Collapses in coal mines killing dozens of workers, spills of toxic coal ash, and gas contamination of water supplies have been constant reminders. The Macondo blowout is another national wake-up call, a sobering plea for action on the greatest environmental challenge of our time: finding a way out of the economic and social model we've built around fossil fuels, and forging a future built instead around the clean energy technologies of tomorrow.

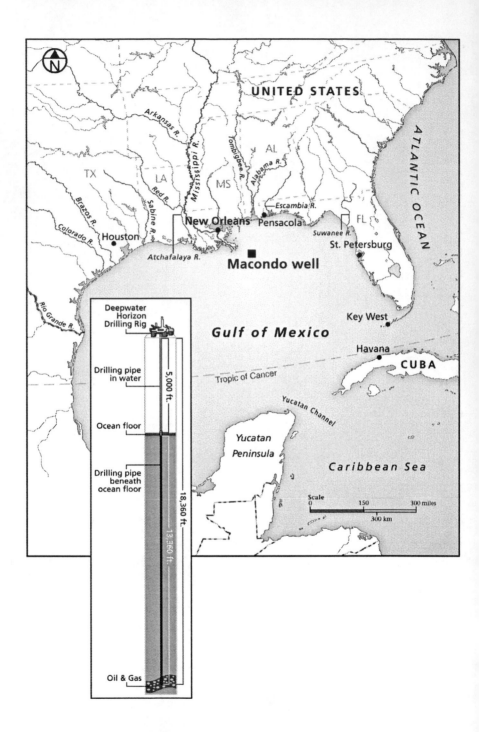

Chapter 1

Blowout

It was just before ten on an April night, and a soft breeze blew over the Gulf of Mexico. For Mike Williams, though, the calm was shattered by the furious hissing of gas, tens of thousands of cubic feet of it, rocketing up from three miles below to the deck of the Deepwater Horizon.

Chief electrician aboard the giant drill rig, forty-eight miles from shore, Williams heard an ear-shattering whine. Diesel engines strong enough to power a jumbo jet began revving out of control, supercharged by the wild gas sucked through air intakes, throwing the generators the engines ran into overdrive. As the machinery ripped past redline, Williams watched in horror as the lights in his workshop, then his computer monitor, flared a blinding magnesium white, then blew out from the electrical overload. He groped in the darkness for the handle of a fire-resistant steel door three inches thick. A bone-rattling explosion blasted the door off its hinges and pinned Williams against the wall.

"And I remember thinking to myself, 'This is it. I'm going to die right here,'" Williams told Scott Pelley of CBS News. Crawling on hands and knees across the debris-littered floor, Williams was rocked by a second explosion before making his way outside, to a deck that looked like a combat zone. A black slurry of seawater, mud and oil had shot up from the well the rig was drilling, gushed like a fountain from the spindly derrick overhead and rained down onto the deck. The derrick was in flames, a raging inferno fueled by gas and oil gushing from the runaway well. Fire was spreading. More explosions rocked the rig—sixty million pounds of steel—like a dinghy hit with a cinder block.

Lifeboats shoved off. Rig workers abandoned ship any way they could in a panicked and chaotic flight from the searing flames. With blood in his eyes from a gash on his head, Williams grabbed a lifejacket.

"I remember closing my eyes and saying a prayer and asking God to tell my wife and my little girl that daddy did everything he could," said Williams. He peered out into the fire-lit water ten stories below and made up his mind to jump. "I made those three steps and I pushed off into the rig. And I fell for what seemed like forever."

Swimming for his life through dark and oily waters and away from flames spreading over the sea, Williams was picked up, pulled into a small boat and taken to the supply ship *Damon B. Bankston* nearby. At eight the next morning, after standing by the burning rig for an all-night vigil in hopes that additional survivors might show, the *Bankston* began the long trip to shore with Williams and other rig workers aboard.

The Coast Guard searched for three days without finding a trace of the men who'd gone missing. Williams and 114 other rig workers survived the April 20 blast. Eleven others would never be found.

After two days burning at sea, the Deepwater Horizon collapsed in mile-deep water over an area of the Gulf called the Mississippi Canyon. As it descended, the rig kinked and tore the steel piping that tethered it to the well, opening a kind of gas and oil volcano that sent nearly two hundred million gallons of crude oil into the ocean over the next three months.

The oil poisoned thousands of square miles of deep ocean, coastal waters, beaches, tidal estuaries and wetlands. It threatened wildlife and habitat from Galveston, Texas to the Florida Keys. It forced the closure of a good portion of Gulf waters to fishing, threw thousands of watermen out of work and put at risk the way of life, and the livelihoods, of millions of Gulf Coast families. On June 15, formally addressing the country from the Oval Office, President Barack Obama said "Already this oil spill is the worst environmental disaster America has ever faced." It would become, in the weeks ahead, the largest peacetime oil spill in the history of the world.

This is the story of the Deepwater Horizon offshore oil disaster, its causes and effects, the role we've all played in this epic debacle, and the road we must travel from here. It is a story of corporate hubris on a colossal and frightening scale. It is a story of short cuts, missed opportunities and warnings ignored. More than a tale of one company, this is the story of an entire industry, the richest and most powerful in the world, that, for far too long, has lived by its own rules, and the rest be damned.

It is also a story, though, of a nation grown so dependent on a single fuel that we've allowed, and in ways enabled, the oil companies to run roughshod over essential safeguards. We've watched, and encouraged, our political leaders, as they've attacked and steadily undermined an entire body of protections put in place to preserve our environment and defend our health.

And we've created tax breaks and other incentives to drive the industry to produce ever more oil, even as doing so has required the oil companies to push the limits of what's safe and reliable and to take extraordinary, and increasing, risks, often without an operative Plan B.

All too often we've turned a blind eye to those risks, part of a broader public failure to understand what we've called on the industry, and the people who work in it, to do on our behalf.

The simple fact is, most of us have little or no idea what is required—from an economic, personal, political, technological, diplomatic, industrial, military or environmental standpoint— to provide this country with the oil we depend on, nearly 800 million gallons every single day, to power our airplanes, cars, trucks and boats with cheap and dependable fuel.

Unless and until we change that, the costs to our Earth will increase, as our demand for this fuel pushes the people and the companies that find and produce it to go to ever-greater lengths, and take on ever-growing risks, to feed our unbridled addiction to oil. That is the great and abiding lesson of the fall of the Deepwater Horizon. That is the cautionary tale we cannot ignore.

On Monday, February the 23rd, 2009, the world awoke to the news that the low-budget film "Slumdog Millionaire" swept eight Oscars at the eighty-first Academy Awards in Los Angeles. In Washington, Congress was locked in a partisan scrum over how to spend hundreds of billions of dollars to try to stimulate the economy out of its worst recession since World War II. And in New Orleans, the U.S. agency that oversees offshore drilling permits in the Gulf of Mexico received a routine, yet fateful, application. The British oil giant BP wanted permission to drill two exploratory wells beneath 4,992 feet of water in the Gulf of Mexico, on a site the company named "Macondo." The drilling was to take no more than one hundred days, or so the applicant stated.

"Under this Exploration Plan, BP Exploration and Production Inc., proposes to drill and abandon two (2) exploratory wells in the Macondo project area," the company stated. "Abandon" is a kind of industry shorthand for the common practice of capping a well once discovered until the company reopens it for production.

For BP, it was a routine, if not perfunctory, filing. The Macondo project, after all, was just one of more than seven hundred sixty drilling leases the company owns the rights to in U.S. Gulf of Mexico waters deeper than one thousand two hundred feet. BP is the single-largest producer of oil from the Gulf of Mexico, where it draws nearly seventeen million gallons a day—or twenty-five percent of the Gulf's total oil output—from fields with names like Thunder Horse, Mad Dog and Atlantis.

To get the oil to shore, BP has built a transportation network it calls "Mardi Gras," an industrial marvel of undersea manifolds, pumps and gathering stations tying together four hundred fifty miles of steel pipelines laced across the ocean floor at depths ranging from four thousand three hundred to seven thousand two hundred feet deep.

The fourth-largest company in the world, by revenues, BP produces one hundred seventy million gallons of oil every day from operations worldwide. With ten percent of that coming from the Gulf of Mexico, the company is deeply invested in the region. And its success finding and producing oil in deepwater wells is one reason the Gulf provides a third of all the domestically produced oil in this country, a percentage steadily on the rise.

BP planners envisioned the Macondo well would be part of that growth. The U.S. Minerals and Management Service approved the drilling proposal within weeks, after reviewing the company's submission, replete with "tentative" drilling schedule and a detailed undersea contour map, depicting in elegantly curved lines the gently sloping seabed of the Mississippi Canyon.

But as seasoned generals sometimes say, the map is not the territory. Macondo soon bore that out. As operator of the lease, BP subcontracted the drilling to the world's largest offshore drilling company, Transocean Ltd. The Swiss-based company—"we are never out of our depth," its literature asserts—is listed on the New York Stock Exchange as simply RIG.

In 2009, Transocean sold $11.6 billion in drilling services worldwide. Its rigs, regarded as ships under maritime law, are among the largest and most advanced in the world. From its fleet of 139 rigs—four times the size of the Canadian Navy—Transocean deploys to every corner of the globe.

In the fall of 2009, though, Deepwater Horizon was booked. That September, the rig drilled what Transocean called the deepest oil and gas well in the world, 35,055 feet, under 4,130 of water, in BP's Tiber well in the Gulf of Mexico. Total distance from the rig to the well: 7.4 miles, more than double the reach of the Macondo well. Transocean began drilling at Macondo October 7, 2009, with another deepwater rig, the Marianas. Just one month into the job, however, a late season hurricane roared through the Gulf and damaged the rig.

It was replaced with Deepwater Horizon, which began drilling on February 6, 2010. Built in the shipyards of Hyundai Heavy Industries, in the South Korean industrial center of Ulsan, the Deepwater Horizon made its maiden voyage to the Gulf of Mexico in 2001, when Transocean took ownership of the $560 million apparatus. A fifth-generation semi-submersible rig, the Deepwater Horizon was a bit like an aircraft carrier with a derrick on the deck instead of an airport. It wasn't anchored over the drill site; it floated on massive pontoons dipping seventy-six feet into the sea. Tethered to the well by a thread of steel pipe twenty-one inches wide, the rig was held in place by what's called dynamic positioning, a sophisticated computer-controlled system that operated eight giant thrusters, kicking out 7,375 horsepower apiece, to keep the rig centered over the well.

A city unto itself, the rig had living quarters for one hundred thirty workers, docking space for supply boats and a helicopter pad. Six diesel engines with a combined force of sixty thousand horsepower drove generators that kicked out forty-two thousand kilowatts, enough electricity to light a good-sized town. It had a deck the size of two football fields and a derrick twenty-four stories high.

It all came at a price: half a million dollars a day in rental fees, and roughly that much again in additional contracting and operating fees. Every day. Seven days a week. For as long as it takes to find oil. At Macondo, that turned out to be a very long time.

Drilling for oil under the best of circumstances is risky and difficult work. First, there is the problem of finding the oil, pinging, probing and pricking the Earth's crust to try to determine the nature of rock formations several miles below, then tying that information to what geologists believe about the paleolithic origins of oil. It is as much art, at times, as science. Nobody gets it right every time. And, with costs typically ranging from $60 million to $100 million to drill an exploratory deepwater well, mistakes take a heavy toll.

"A dry hole and you feel like jumping out of the window," BP's vice president for Gulf of Mexico production, David Rainey, told Tom Bower, author of the book *Oil: Money, Politics and*

Power in the 21st Century. The emotions, said Rainey, "are indescribable."

Even when the geology is right, the timing can be off. Oil can be unusable. It can leach away through porous rock. It can seem to disappear. In 1983, the Sohio Oil Company—then majority-owned by BP, which later bought the company outright—spent $1.6 billion to drill in the Beaufort Sea, off the coast of Alaska, on a site geologists were certain held at least forty billion gallons of oil. There was, indeed, liquid trapped deep beneath impermeable stone. It was saltwater.

"We drilled in the right place, we were simply thirty million years too late," Richard Bray, president of Sohio's production company, told reporters at the time, energy analyst Daniel Yergin wrote in *The Prize*, his Pulitzer-winning history of the oil industry.

When the geologists do get it right, and the drillers show up in the right epoch, there remains the matter of poking a hole into rock hard enough to hold hot oil that's been seething for millions of years beneath the incalculable weight of sediment and stone often several miles thick. Drilling in deep oceans is all that and more, adding uncertainties and harsh conditions that make deepwater drilling something like operating in outer space.

Opening a deepwater well in the Gulf of Mexico is like popping the cork on a bottle of oil at two hundred fifty degrees Fahrenheit and under as much as eighteen thousand pounds per square inch of pressure. Imagine nine Volkswagens stacked up on a postage stamp and you get the general idea. Part of the trick is to open the bottle at all. The other part is to keep it from blowing up in your hand. Drillers use a term of art called a "well control event." Put a bit more clearly, it means the well's flirting with blowout. Pressures can exceed expectations. Rock formations can crumble and crack. Drilling operations can falter. The well itself can fail.

From the time drilling began in February until April 6, Transocean's records show, there were at least four such events at Macondo. Each was contained, but the project had shown itself to be unpredictable enough that it became known as a "nightmare well which has everyone all over the place," as BP drilling engineer Brian Morel wrote in an April 14 email to a colleague.

Other rig workers echoed the theme. "There was always like an ominous feeling," rig survivor Daniel Barron III told CNN's Anderson Cooper. "This well did not want to be drilled."

Still others spoke even more starkly about the fate of Macondo, the name, as it happens, of the snakebitten fictional town wiped out by a hurricane in the 1967 novel *One Hundred Years of Solitude* by Colombian author Gabriel García Márquez.

Adam Weise, a twenty-four-year-old rig worker from Yorktown, Texas, was nearing the end of the final shift of his three-week rotation on Deepwater Horizon the night of April 20, when a supervisor asked for help in a pump room. That's the last place Adam is known to have been seen alive. His girlfriend, Cindy Shelton, told the *Houston Chronicle* newspaper that Andrew had been calling her from the rig before and after every shift—unusual for him—invariably expressing frustration over the troubled project.

"Everything that could go wrong was going wrong," she said. "Every time he'd call me, he'd say, 'this is a well from hell.'"

A civil engineer from the Texas oil town of Wichita Falls, Rex Tillerson is no critic of the petroleum industry. Since joining Exxon as a production engineer in 1975, he's worked his way to the top, becoming chairman and chief executive officer in 2006.

An unabashed defender of Exxon's record-breaking $45.2-billion profit in 2008—great risk warrants handsome rewards—Tillerson once dismissed alternative fuels like ethanol as "moonshine" and, in a 2008 interview with *The New York Times*, blamed "election cycle" politics for the lack of a long-term U.S. energy policy. On the subject of the Macondo well Tillerson is equally direct.

"It appears clear to me that a number of design standards that I would consider to be the industry norm were not followed," Tillerson testified in a June 17, 2010, congressional hearing on the Deepwater Horizon disaster. "We would not have drilled the well they way they did."

Responding to questions from members of the House subcommittee on Oversight and Investigations, Tillerson took the Macondo operation apart. He criticized the well's design, in particular the outside pipe, called casing, reaching down from the rig to the well. A more popular design, he said, would have provided greater protection against a blowout, at marginally higher cost. He took issue with the formulation of the cement used to seal the casing and hold it in place. Moreover, he said, operators should have tested the cement once it hardened, before pumping out the industrial drilling mud used to counter the well's pressure. A steel collar should have been installed to lock down a crucial juncture of pipe. And, finally, he delivered this operational broadside: "There were clearly a lot of indications of problems with this well going on for some period of time leading up to the final loss of control. How those were dealt with, and why they weren't dealt with differently, I don't know."

Tillerson, moreover, was not alone. "We wouldn't have drilled a well that way," James Mulva, chairman and chief executive officer of ConocoPhillips, told the subcommittee. "It's not a well that we would have drilled with that mechanical setup," echoed Marvin Odum, president of the Shell Oil Company. "And there are operational concerns."

John Watson, chairman and chief executive officer of the Chevron Corporation, said the casing design and mechanical barriers to prevent a blowout "appear to be different than what we would use."

In some ways it wasn't surprising for BP's rivals to assail the underpinnings of a disaster that had become the Internet banner ad for what's wrong with Big Oil. In attacking BP, its competitors were defending the industry. Still, the unanimity of the criticism—of BP and of its chief executive officer, Tony Hayward, scheduled to be replaced in October—reflected a widening consensus that Macondo was a project riddled with faulty equipment, flawed decision-making and operational failures of multiple stripes.

"My first thought was, 'Man, did they hang him out to dry,'" said Julius Langlinais, a former offshore drilling engineer who is now a professor of petroleum engineering at Louisiana State University in Baton Rouge. "But, to a certain extent, I have to agree with them."

As of this writing it isn't yet known what caused the blowout at Macondo, though theories and clues abound. The matter is under investigation by the National Commission on the BP Deepwater Horizon Oil Spill and Offshore Drilling, appointed by President Obama. It's also being probed by the U.S. Coast Guard; the freshly-minted Bureau of Ocean Energy Management, Regulation and Enforcement that Obama set up to replace the discredited Minerals Management Service; the U.S. Justice Department and several congressional committees, among others. BP and key contractors are conducting internal inquiries and reviews, all of which will become central to lawsuits likely to last years, with billions of dollars in liability claims at stake.

This much is known for certain: on the night of April 20, after repeated warning signs went unheeded, pressures inside the Macondo exploratory well overwhelmed the system and blew out the well, spewing out oil and natural gas, which ignited, exploded and eventually burned down the rig. It happened just as the drilling crew was in the final stages of capping the well and putting it to bed, until its production phase.

There was a failure of either the system of steel pipes and connectors leading out of the well or the cement poured to seal those pipes and hold them in place. It's possible both the piping and the cement failed. At the same time, equipment that was supposed to shut down a blowout or mitigate its effects either didn't function or wasn't in place.

And a series of decisions made by key players, from the well's designers to members of the rig's operating crew, appears to have stacked the odds against a safe and reliable well. On board the Deepwater Horizon, "drill baby drill" was no Tea Party bumper sticker: it was a cultural mandate fueled by the corporate imperative to seal the well and move on to drill the next hole in the seabed.

"Run it, break it, fix it, that's how they work," one worker told an investigator with Lloyd's Register Group, which helps companies manage maritime risk, during inspections a month before the blowout. Workers surveyed said production demands often took precedence over safety and maintenance concerns, and only about half the workers said they felt comfortable reporting risky conditions without fear of reprisal, according to *The New York Times,* which managed to get copies of a pair of confidential reports that detailed the findings.

It isn't yet known whether, or to what extent, that kind of atmosphere might have contributed to the Macondo blowout. Investigators, though, quickly seized on the question of whether a well that was costing BP a million dollars a day was driving the company to take short cuts that ultimately led to disaster.

"BP cut corner after corner to save a million dollars here, a few hours or days there," concluded House Energy and Commerce Committee Chairman Henry Waxman (D-Calif.). "And now the whole Gulf Coast is paying the price."

Tony Hayward rejected that characterization. "We have focused like a laser on safe and reliable operations—that is a fact—every day," Hayward, a geologist with twenty-eight years at BP, told Waxman and other members of the subcommittee on Oversight and Investigations at a hearing June 17, 2010.

At the same time, said Hayward, BP's own investigation of the blowout raised a host of concerns early on that the company was continuing to assess. "I think we will find that this was an incredibly complicated set of events with individual decisions and equipment failures that led to a very complicated industrial accident," Robert Dudley, slated to replace Hayward as BP's chief executive officer in October, told *The New York Times* in July. "I don't accept, and have not witnessed, this cutting of corners and the sacrifice of safety to drive results."

And yet, the unadorned Macondo record reveals a stunning litany of irregularities, missed signals and compromises apparently driven by schedule and cost. It may turn out that no single individual action, decision or omission was determinative. Taken together, though, the long train of apparent deficiencies and evident missteps set the stage for catastrophe that faulty backup systems failed to prevent, congressional members from both parties said.

"This rig was forty days behind schedule and millions of dollars over [budget], and there was a lot of pressure to finish the job. And there was nobody on that rig whose job was to make sure that they made the safe decisions," Representative Joe Barton of Texas, the ranking Republican on the House Energy and Commerce Committee, said at the June 15 hearing. "And so, when you start making decision after decision after decision that is not, in and of itself, a bad decision, but cumulatively minimizes safety, eventually you reach a critical mass and you have an accident that happens."

That seems a fair description of the mosaic that emerges through sworn testimony, documents and statements from regulators, BP officials, rig workers, contractors, engineers and scholars that have appeared before various investigatory bodies to provide a grim anatomy of a catastrophe in the making.

When drillers bore into the ground in search of oil, they line the hole with steel pipe, called casing, then pump cement around the outside of the pipe to seal it and hold it in place. Otherwise, gas and oil under pressure can flow between the casing and the rock that surrounds it, or up through the piping itself. That can happen also if the cement or casing fails. The day before the Macondo blowout, workers installed a final section of casing in the well, completing an operation investigators are studying for possible clues as to what might have caused, or contributed to, the disaster.

From the rig to the bottom of the Macondo well was 18,360 feet. The first mile of that was water; the other two and a half miles were drilled into rock and sediment from the sea bottom down. As workers prepared to seal the well, there was casing lining the hole for all but the bottom one thousand two hundred feet. To attach that final portion of pipe, BP had two options. It could run what is variously called a single string, or long string, of casing, which runs from the well head, at the ocean floor, to the bottom of the hole, or run a liner, which is attached to the base of the existing pipe. If a liner is run, then a long string of pipe is then run from the well head to the top of the installed liner, connected with a steel seal, then cemented into place. Imagine running a narrow straw inside a pair of wider straws and you get the general idea.

In a series of internal BP documents and emails obtained and made public by the House Energy and Commerce Committee, the company laid out and discussed the pros and cons of each approach. The single string was the faster and cheaper way to go, but it was potentially riskier. Here's why.

The rock in the well itself, a porous sandstone reservoir of oil and gas, was under pressure and brittle in places. That meant that a sound cement job wasn't assured, because when rock crumbles and cracks, it can get into the cement, weakening it, clog up channels where the cement is supposed to flow or create rifts that will draw the cement away.

"Cement simulations indicate it is unlikely to be a successful cement job due to formation breakdown," states a BP "Forward Plan Review" outlining the final casing options.

If the cement failed to seal the casing properly, oil and gas would have to penetrate additional barriers—seals, essentially, in the casing—to blow out the well. The single string approach, due to its design, provided two barriers to a blowout, according to documents provided to the committee by the Halliburton Company, the oil services company that BP hired to cement the well.

The costlier alternative—using the tieback liner—would

have provided four barriers. That design was, in essence, the more blowout resistant of the two. That option, though, would have cost between $7 million and $10 million more than the long string approach and would have taken at least three days longer to install, BP estimated. The company went with the lower-cost option.

"The conclusion I draw from these documents is that BP used a more dangerous well design to save $7 million," Waxman told Hayward. "If you made mistakes, the consequences of those would be catastrophic, and, in fact, have turned out to be catastrophic," said Waxman.

"It was approved," by federal regulators, Hayward countered. "The long string is not an unusual well design in the Gulf of Mexico."

Halliburton challenged that, telling House staffers that the long string design is used in no more than ten percent of Gulf of Mexico deepwater wells, the tieback being the preferred option by far, as the other oil company executives testified.

BP internal documents suggest that cost and expediency were factors. The long string "saves a lot of time . . . at least three days," BP drilling engineer Brian Morel wrote in a March 25 email to Allison Crane, materials management coordinator for BP's Gulf of Mexico Deepwater Exploration Unit. In a separate document, BP estimated that using a tieback liner would "add an additional $7-$10 MM to the completion cost," and concluded that "we should be able to achieve a successful primary cement job on the long string."

A sound cement job, though, also depends on proper positioning of the pipe being cemented in the hole. One way that's assured is through the use of centralizers, spring-loaded attachments that ring the outside of the pipe as it's lowered into the hole. Once the pipe is in place, the centralizers push against the surrounding rock to hold the pipe more or less centered in the hole. Otherwise, pipe tends to lean against one side or another of the hole. The proper flow of cement, and thus the formation of a snug jacket around the pipe, is impeded. The result can be weak or open passages that gas and oil can blow through.

On April 15, the day after referring to Macondo as a "nightmare well," Morel, the BP drilling engineer, informed Halliburton account representative Jesse Gagliano that BP would use the six centralizers already aboard the rig to secure the final one thousand two hundred feet of casing. Gagliano ran a computer analysis that showed it would take twenty-one centralizers to do the job properly.

Gagliano shared his analysis with Morel. Even using ten centralizers, Galiano warned, would result in a "moderate" problem with gas finding its way up the side of the pipe, while BP's design would result in "severe" gas flow problems. "It's too late to get any more product on the rig," Morel replied.

There was, though, another option. On April 16, Gregory Walz, drilling engineering team leader for BP's Gulf of Mexico drilling and completions unit, tracked down fifteen centralizers in Houston that could be flown out to the rig the next morning. "My understanding is that there is no incremental cost with the flight because they are combining the planned flights they already had," Walz wrote in an email explaining the option to BP's drill team leader, John Guide.

Walz contended in his email that, since BP was using the single string design, it should respect the analysis of the cement professionals at Halliburton as to how many centralizers it would take to do the job right. "We need to honor the modeling to be consistent with our previous decisions to go with the long string," he wrote. He drove home the point by reminding Guide of a well the company drilled earlier in the Gulf, on its Atlantis field, that apparently went awry.

"I wanted to make sure that we did not have a repeat of the last Atlantis job with questionable centralizers going into the hole," Walz wrote. "I do not like or want to disrupt your operations and I am a full believer that the rig needs only one Team Leader," Walz added by way of apology. "I know the planning has been lagging behind the operations and I have to turn this around." If Walz was anticipating resistance from Guide, his concerns were quickly affirmed. "I do not like this," Guide shot back in reply.

To a foreman working on a deadline, centralizers can seem an unwelcomed burden. They sometimes hang up pipe as it's lowered in ways that require it to be pulled back up again. "It's jewelry on that pipe, and these old guys don't like that," explained Langlinais, the LSU petroleum engineering professor. "If anything goes wrong with the centralizers, you've got to start over."

Morel moved forward with plans for just six centralizers. On April 16, he emailed a colleague, Brett Cocales, "I don't understand Jesse's centralizer requirements." Internal assessments, Morel wrote, assured him that six would suffice if spread out properly. "Even if the hole is perfectly straight, a straight piece of pipe even in tension will not seek the perfect center of the hole unless it has something to centralize it," Cocales, a BP operations drilling engineer, explained to Morel in reply. "But, who cares, it's done, end of story, will probably be fine and we'll get a good cement job."

The following day, Macondo's drilling was completed. Over the next three days, the final pipe went down and was cemented in place. If the cement didn't form a solid collar, Cocales wrote, it could be patched through a common remedial operation referred to as a "squeeze." Even that option, though, was later scuttled by BP.

A "squeeze" is when cement is injected between pipe and rock to fill in gaps that might occur in a cementing job. To find those gaps requires that the job be tested with special equipment dropped down the pipe on a wire after the cement hardens. The equipment emits powerful sound waves, then processes the echo, a bit like the way sonar works. The test, called a cement bond log, tells technicians whether there's solid cement around the pipe or something softer—mud, sandstone, loose rock or the like—that oil or gas might blow past. If more cement is needed anywhere, technicians send down a tool to drill a small hole in the pipe, through which additional cement is injected, or "squeezed."

Industry experts consider the cement bond log an essential backstop, especially for the kind of single string casing BP used at Macondo, because that design depends so heavily on a good cement job to prevent a blowout. "If the cement is to be relied upon as an effective barrier, the well owner must perform a cement evaluation as part of a comprehensive systems integrity test," Halliburton's vice president of cementing, Tommy Roth, told staff members of the House Committee on Energy and Commerce during a June 3 briefing.

Indeed, late on the night of April 18, BP flew a special crew out to the rig to perform the cement bond log. The team was with the Houston-based oilfield services giant, Schlumberger Ltd. The cement log, though, was never performed.

A Schlumberger timeline shows that at 7 AM on April 20, BP told team members they could go home. Four hours later they boarded a BP helicopter and flew off the rig. That was eleven hours before the blowout.

BP saved $128,000 by scuttling the cement bond log, but at what ultimate cost? The test might have identified cement flaws, flaws that could have caused, or contributed to, the blowout. "That may have saved them had they run the cement bond log," said Langlinais. "That would have identified the problem at a point in time where they could have done something about it. They could have recovered," he said. "So the cement bond log decision was sort of this point of no return."

Not that operations went smoothly from there. In the hour before the blowout, there were at least three signs of trouble. In each case—one fifty-one minutes before the explosion, another forty-one minutes before and a final one eighteen minutes ahead of the blast—pressure forced saltwater, oil, mud and gas out of the well and up the pipe toward the rig, according to an internal BP report. At any one of those points, the well could have been shut down, through the use of industrial drilling mud. But that didn't happen.

A mix of clay, chemicals and heavy minerals, drill mud is roughly twice as heavy as oil. Drill operators use it to dampen a well's pressure as the well is being sealed. The operation is a tightrope walk. Too much downward pressure can shatter the soft rock—in this case, sandstone—that forms the sponge-like reservoir that holds the oil in the well. The pressure surges and fluid flows in the hour before the blowout showed how much trouble well operators were having keeping pressures balanced. Eventually, they lost control.

There was still a final backup, a blowout preventer, the well's ultimate failsafe system: a five-story high, $15 million piece of equipment near the ocean floor. A stack of pipes, valves and heavy blades, a blowout preventer is meant to crush, shear and seal the pipe in the event disaster strikes.

For reasons investigators have yet to determine, the blowout preventer failed in a number of ways. Not only did it fail to shut down the well, it also failed to perform as designed and detach from the riser—the long pipe leading up from the well to the rig—when the Deepwater Horizon sank. The blowout preventer's "deadman" switch, meant to shut down the well when communications with the rig is lost, never engaged. Even after the rig sank, the blowout preventer didn't respond when remotely operated vehicles were used to try to manually engage the unit undersea.

It could be some time before it's known what went wrong with the blowout preventer. After all, it's sitting on the bottom of the ocean and weighs 325 tons. What is known is that the blowout preventer showed signs of malfunctioning in a number of ways for weeks, even months, before the disaster ensued. The real question is, why wasn't it fixed?

The Deepwater Horizon's blowout preventer had two control pods—a primary control and a backup. One had a dead battery and had been malfunctioning long before the blowout. And Ronald Sepulvado, the BP rig foreman up until April 16, testified July 20 that he had reported a hydraulic leak in the pod months before the accident. When a blowout preventer control pod malfunctions, federal regulations require drilling operations to be suspended until it's fixed. Sepulvado said he reported the problem to the BP team leader—John Guide—figuring that was enough.

"I assumed everything was okay, because I reported it to the team leader, and he should have reported it to MMS," Sepulvado said in testimony in New Orleans before the Marine Board, a joint investigatory body made up of representatives from the U.S. Coast Guard and the Bureau of Ocean Energy, Management and Regulation—formerly the Minerals Management Service.

On April 6, a second blowout preventer problem emerged, when a rig worker noticed clumps of rubber the size of his fist in drill mud that returned to the surface after the drill pipe was accidentally pulled through the inside of a large doughnut-shaped gasket in the center of the blowout preventer. The gasket is essential, part of what seals the well hole when drilling pipe come out.

Rig workers shrugged off the red flag, attributing the rubber chunks to normal wear on the seal, according to a Transocean document. Four days later the blowout preventer passed a pressure test, but the damaged seal could have compromised the integrity of the test. It could also account for similar pressure discrepancies in the hours leading up to the blowout.

In fact, the term blowout preventer is a misnomer. The equipment has contributed to accidents or failed outright in at least fourteen accidents, mostly since 2005, according to an Associated Press review of federal safety records for offshore oil operations. A federal report in 1999 identified 117 blowout preventer failures in the previous year. Not all resulted in oil spills, but a number of them did.

A common problem is failure in a key component called a blind shear ram, designed to slam a leaking pipe with a pair of blades at up to a million pounds of pressure, cut it and seal it, shutting down a blowout. That component didn't work at Macondo. Investigators want to find out why not. The Deepwater Horizon's blowout preventer, moreover, had only one blind shear ram. Newer blowout preventers have a back up, because studies have shown that the blind shear rams themselves fail almost as often as they work.

Five hours before the blowout, there was an unexpected loss of fluid in the riser pipe—a sign that the blowout preventer may have been leaking, BP's preliminary review of the incident found. Three hours after that, there were indications that there was three times as much fluid in the piping system as could be accounted for, suggesting that oil could be surging up from the well. A related set of pressure issues pointed to what a BP investigator called "a very large abnormality" in the well system. Rather than shut it down, however, the crew monitored the system, taking note of the three additional trouble signs in the hour before the blowout. Why didn't the well operators use more drilling mud to tamp down, or even shut down, the well, given the series of problems they were having keeping its pressure in check?

At nine minutes before ten, the well blew out, the rig exploded, and communications with well conditions went dark.

With investigations pending, and so much still unknown, conclusions are premature. The House Energy and Commerce Committee, though, issued a scathing assessment of its preliminary findings in June. "Time after time, it appears that BP made decisions that increased the risk of a blowout to save the company time or expense," the committee wrote in a June 14 report. "If this is what happened, BP's carelessness and complacency have inflicted a heavy toll on the Gulf, its inhabitants, and the workers on the rig."

On a rainy Tuesday, five weeks to the day after the Deepwater Horizon disaster, hundreds gathered in the delta capital of Jackson, Mississippi, to commemorate the men who lost their lives on the rig. At a makeshift altar in the Jackson Convention

Complex, a simple white cross was surrounded by eleven bronze-colored hardhats, one for each of the men who died. In a video tribute, their names and faces flashed by on a screen: Donald Clark, Roy Kemp, Stephen Curtis, Gordon Jones and Blair Manuel, from Louisiana. Two Texans—Jason Anderson and Adam Weise. Dale Burkeen, Karl Kleppinger, Dewey Revette and Shane Roshto, from Mississippi. "Missing You," country music star Trace Adkins' paean to offshore rig workers, played in the background.

"I've worked through broken drills and busted hands, weeks without seeing dry land," sang Adkins, who made his living for six years on offshore rigs in the Gulf before hitting pay dirt as a singer. "I'll work as hard as any man. But, until I'm home with you again, the toughest thing out here that I go through is missing you."

Now the longing knows no end for people like Arleen Weise, one of twenty-one of Adam's family and friends who flew from Yorktown, Texas, to attend the memorial service in Jackson, organized by Transocean. "These last few days it has hit me that my son is never coming back to me," she told the *Houston Chronicle*. "I'm not holding it together," she said. "I keep seeming to be more of a mess."

A ship's bell tolled eleven times to mark the lives of the men. Each family received a hard hat and a book called *The Beacon*, with pictures and brief remembrances for the fallen. There were prayers, hugs and tears. There were no caskets or urns, no remains to inter; all lost to the fire and the sea.

"There will never be complete closure," Burkeen's uncle Aaron Bryan told the Associated Press. "We don't have the body to see."

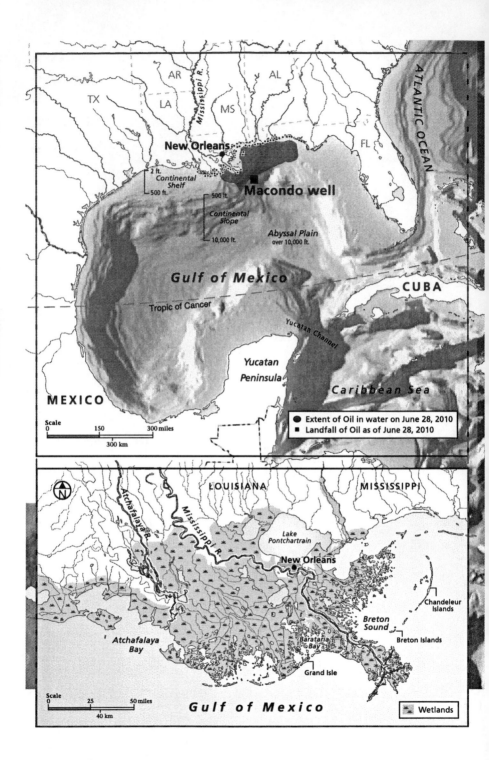

Scale
0   150   300 miles
300 km

● Extent of Oil in water on June 28, 2010
■ Landfall of Oil as of June 28, 2010

Scale
0   25   50 miles
40 km

🌿 Wetlands

Chapter 2

Oil in the Water

Derrick Evans filled his lungs with pungent delta air, squinted into the sunlight glaring off the brackish waters of Grand Bayou, and gazed out toward the Gulf of Mexico. A mile away lay the leading edge of millions of gallons of oil, its toxic film bearing down like some predatory pall upon the verdant wetlands an hour's drive south of New Orleans.

"The reason there's so much oil out there is because there has always been so much life here," said Evans, a history teacher from Biloxi, Mississippi. The origins of petroleum reach back tens of millions of years. "The same forces that made this America's wetlands made the area America's carbon graveyard."

Nature and geologic time have combined to create a modern paradox in the Gulf of Mexico. Home to some of the richest ocean and coastal habitat anywhere on Earth, it's become our national filling station. A complex and interwoven web of deep ocean, coastal waters, wetlands, estuaries, beaches, barrier islands, bayous and bays, the Gulf is a natural wonderland, an ecological treasure chest opening itself up to thousands of species of birds, wildlife, marine and aquatic life and an endless array of plants. From out of its fertile waters and shores come seventy percent of the shrimp and oysters produced in this country, along with hundreds of millions of pounds each year of red snapper, blue crabs, yellowfin grouper and other seafood. It is also the source of a third of the oil produced in the United States, 1.7 million barrels a day, from offshore wells that pepper the sea.

In some ways this is a tale written in stone. For the sediment-laden rivers that have nourished and renewed life in the Gulf for eons have also carried the sand, clay and silt that formed a giant crucible for turning the remains of ancient plants and animals into vast amounts of oil. "The key is that you have so much sediment, over such a long period of time" explained Jeff Williams, scientist emeritus with the U.S. Geological Survey. "That's really the bottom line as to why the Gulf of Mexico is such an attractive place for oil and gas," said Williams, a coastal marine geologist who spent thirty years researching the Gulf. "It's also just very conducive to a lot of biologic production-both for the wetlands and the fisheries."

The Gulf of Mexico began opening up as a rift along the crest of South America, as that continent began pulling away from North America two hundred million years ago. Over time the cleavage was widened, contorted, twisted and pulled into what it is today, an oblong basin more than twice the size of Texas with a shoreline stretching three thousand six hundred miles from Cape Sable, Florida to the tip of the Yucatan. Sometimes called the "American Mediterranean," the Gulf mixes the waters of the Atlantic Ocean and the Caribbean, through currents that feed the Gulf Stream.

For the past forty million years, the Gulf also has been watered by the snowmelt, groundwater, glacial runoff and rain rolling off the broad middle of North America and into the Mississippi River. Fed by major rivers and lesser tributaries from the Rocky Mountains to the Appalachians, the Mississippi drains forty percent of the United States, and then pours itself into the Gulf. En route, the river brings with it the fine silts washing out of forests and fields and the rich residue of geologic change, the coarse sand and chips and flakes of stone running off the ever-changing face of the land. The river, in a sense, skims up the mineral and organic fruits of the continent, and delivers them to the Gulf, along with some other of the country's most storied waterways, including the Rio Grande, the Colorado, the Escambia and the Suwannee rivers.

Along the Gulf's fringes, the fertile sediments these rivers bring have formed one of the largest and most complex coastal delta plains anywhere in the world. The mouth of the Mississippi has wandered over time from Beaumont, Texas to as far east as the Chandeleur Islands. "It slops around at the Louisiana end like a garden hose geologically," said Roger Anderson, a Columbia University geophysicist. In the process, the river has built up forty percent of our nation's tidal wetlands in Louisiana alone, enough to cover the state of Maryland. The wetlands border brackish estuaries and bays, creating a vital buffer against storms and hurricanes, as well as essential habitat for thousands of species of wildlife, fish and waterfowl. Beaches and barrier islands take the brunt of the force of currents and waves, providing nesting grounds for turtles and pelicans, shelter for young shrimp and speckled trout and safe harbor for millions of migratory birds. Nursery, sanctuary and breadbasket, the delta plain is the foundation of life in the Gulf.

"Ninety-seven percent, by weight, of the commercial fish and shellfish landings from the Gulf of Mexico are species that depend on estuaries and adjacent wetlands for some point of their life cycle," said David Westerholm, director of the Office of Response and Restoration, which oversees efforts to combat the environmental impact of the oil spill for the National Oceanic and Atmospheric Administration. The area is also one of the country's top destinations for recreational fishermen, who come to catch everything from redfish to blue marlin.

"Landings from the coastal zone in Louisiana alone make up nearly one-third of the fish harvested in the continental United States," Westerholm told a House Natural Resources subcommittee June 10, 2010. "In such an incredibly productive area, important habitat covers nearly every part of the ecosystem."

Swamps of bald cypress and tupelo gum, barrier islands of wax myrtles and black mangroves, thick beds of seagrass harboring young crabs, coral reefs, beaches and tidal estuaries are the keystone attractions in more than a dozen state and national parks and protected areas along the Gulf Coast from Texas to the Florida Keys.

President Theodore Roosevelt set up the Breton National Wildlife Refuge in 1904 to protect Breton Island and the nearby chain of Chandeleur Islands, important nesting grounds for royal terns, loggerhead sea turtles and piping plovers. It is the second-oldest of the country's five hundred fifty national wildlife refuges.

West Indian manatees, wild turkeys and Gulf sturgeon thrive in the tidal flats and pine forests of Florida's Lower Suwannee National Wildlife Refuge. Armadillos, great blue herons and ruby-throated hummingbirds abound in the sand dunes and sea oats of the Bon Secour National Wildlife Refuge in Alabama. And Louisiana's Atchafalaya National Wildlife Refuge takes in part of the country's largest bottomland hardwood swamp, home to American alligators, Florida panthers and Louisiana black bears.

The same forces that have made the region a showcase of natural diversity, and the sustainer of life in the Gulf, have also left rich and valuable deposits along the Continental Shelf—a gently sloping underwater plain ranging from about sixty to five hundred feet in depth and extending between ten and 150 miles from shore—and in the deeper water further offshore. There, sediments borne by the same great river systems that created the modern delta plain have piled up more than ten miles deep in places, hardening with pressure and time into sandstone and shale.

The ocean has added its bit. Over many scores of millions of years, the remains of dead marine life—from tiny plankton and algae to the largest whales—fell to the sea bottom and became part of the mix, as organic-rich lignite, lime mudstone and marl.

The Gulf shoreline, meanwhile, has advanced and retreated hundreds of miles over time, with the rise and fall of sea level. Those changes left behind rich deposits of plant and animal remains, as well as broad sheets of salt. Some of the salt layers run more than a mile thick, along a band sweeping from Texas across Louisiana, extending more than two hundred miles offshore into some of the deepest waters of the Gulf.

This sandwich of sub-sea sediment and salt, a stratified layer cake of mineral and stone, has made the Gulf of Mexico one of the best places anywhere in the world for turning life into oil. Heated and pressed for millions of years, ancient plant and animal remains were reduced to their hydrocarbon basics—natural gas and oil—then stored in great reservoirs of porous sandstone and trapped there by an impermeable cap of salt or hard shale.

The result is an estimated forty billion barrels of oil in American waters in the Gulf, according to the latest government and industry estimates. At $75 a barrel, that oil is worth $3 trillion. The Gulf is still making oil today, of course, but it'll be a few million years before that's ready. In the first two months of 2010, offshore wells in the Gulf of Mexico pumped 1.7 million barrels of oil a day, up seventy percent since 1996, according to the Energy Information Administration, the analytics and statistics wing of the Department of Energy. All told, Gulf offshore oil makes up thrity-one percent of U.S. domestic production, as much as comes from onshore wells in Texas and Alaska combined.

"It's always been here," said Evans, "a kind of burial site with all kinds of living history since the beginning of time, at rest and at peace—until we decided to go and disturb it."

It was just after breakfast on a morning in May along the edge of Barataria Bay, a broad bowl-shaped estuary teeming each spring with young shrimp and speckled trout. The air was already steamy, the aluminum boat warm to the touch, as Byron Encalade glided slowly past plywood cabins with sagging roofs and names like Wahoo, Dixie and Amazing Grace. Reaching into a wire bin, he pulled out a charcoal-colored oyster the size of his fist. It was raised after Hurricane Katrina roared through five years before, ripping at the tidal shallows and leaving oystermen like Encalade to start over again from scratch.

"This is what we've been waiting for since Katrina," he said. "Now it's all about to be destroyed." He tossed the oyster back in the bin, gazed out across a narrow channel and shook his head. "Goin' down the drain."

When two hundred million gallons of crude oil spewed out of BP's Macondo well, some of it wound up fifty miles away, in oyster beds Encalade has worked since he was a boy. President of the Louisiana Oystermen's Association, he's a third-generation waterman who leads a crew of eight: his brother, two nephews and five cousins. They measure their wealth in burlap bags, oysters at $28 a sack, a marsh grassroots part of the seafood industry that, in better times, brings Louisiana $2.4 billion a year. This year, though, is different, for Louisiana and its watermen. "Our oyster beds are dead," Encalade said in late July. "We're talking about our lives here now."

Weeks after the Macondo blowout, state and federal wildlife authorities began closing Gulf Coast fishing grounds from the mouth of the Mississippi to the Pensacola Bay, out of concern that oil would contaminate seafood. Oysters, for example, are filter feeders. They suck in water and filter out nutrients, so they're easily and directly contaminated by pollutants. By June 2, nearly ninety thousand square miles of the Gulf of Mexico—thirty-seven percent of U.S.-controlled waters there—were off-limits to fishing. The closings idled thousands of fishermen and boats from Texas to Florida, costing watermen hundreds of millions of dollars in lost income.

"Birds don't fly, fish don't swim and fishermen can't make a living," Charlotte Randolph, president of Laforche Parish, a Gulf Coast county in southern Louisiana, told the National Commission on the BP Oil Disaster and Offshore Drilling during a July 2010 hearing.

It isn't only this year's catch that's at stake. Shrimp larvae shelter in the estuaries until they're big enough to head to deeper coastal waters. Pollution that lingers in the wetlands threatens shrimp when they're young, weak and vulnerable. Damage to this year's hatchlings will impact populations for years to come. Oyster beds are leased by watermen and often managed by the same family for decades. Typically they are rotated, much like farm crops, with beds being harvested then seeded with oyster larvae, or "spat," that take between three and five years to reach shucking size. With beds wiped out entirely, though, it is not clear how watermen like Encalade can hang on.

"After Katrina, forty percent of our fishermen exited the business," said Robin Barnes, executive vice president for Greater New Orleans Inc., a regional economic alliance. "You don't have people who have savings. You have zero slack capacity to survive something like this. A lot of these guys will not survive the closing of the whole season," she said, to say nothing of the next four or five years.

"I don't even want to think about it," said Maurice Phillips, who leases 250 acres of oyster beds near Barataria Bay. "We don't know how to do anything but shrimp, catch oysters and trap," said Phillips. "We hope that some way, somehow, the Good Lord will open up a way to provide."

There are questions as well about how soon oyster beds will be suitable for seeding. Two decades after the *Exxon Valdez* supertanker spilled nearly eleven million gallons of oil in to Prince William Sound, oil is still being found on the rock-ridden beaches of the Alaskan shore. Fiddler crabs near West Falmouth, Massachusetts still display peculiar behavior thirty years after the barge *Florida* ran aground, spilling nearly eight million gallons of oil nearby. The history of these and other spills has shown that oil can lurk just underground, where it continues to harm plant and animal life for decades.

"Long-term issues of restoring the environment and the habitats and stuff will be years," U.S. Coast Guard Vice Admiral Thad Allen, national incident commander for the federal response to the spill, told reporters at the White House June 7. With so much marine and aquatic life dependent on clean Gulf waters and healthy coastal habitat for reproduction and early maturation, enduring damage to the populations of many species is highly likely. In some cases, the damage could prove terminal, particularly for species already fighting for their survival in the face of habitat loss and overfishing.

The Gulf is the spawning area for the entire population of Western Atlantic bluefin tuna, a princely fish that can live for two decades, grow to as much as fourteen hundred pounds and swim through waters as far north as Newfoundland at speeds of up to fifty miles per hour. Like all creatures, though, they start off small and vulnerable. At eight years old, they begin to spawn and half don't attain the ability to reproduce until they reach twelve. The oil spill occurred just as the two-month spawning season began. And the oil poisoned the waters just as fragile eggs and larvae were set adrift in floating sargassum grass, acres upon acres of which were destroyed by oil.

Even before the spill, stocks of these fish in U.S. waters had plummeted by more than seventy-three percent since 1975, mostly due to overfishing, according to the National Oceanic and Atmospheric Administration's fisheries service. With oil threatening to disrupt, if not devastate, their ability to reproduce, these magnificent fish are perilously close to a complete population collapse. The bluefin tuna is on the brink of extinction. The Center for Biological Diversity, an environmental advocacy group based in San Francisco, has petitioned the Commerce Department's National Marine Fisheries Service to list the fish as an endangered species. My organization, the Natural Resources Defense Council, supports this petition and we hope it prevails. We can't allow this magnificent creature to disappear from the face of the earth.

In November, 2009, another iconic Gulf Coast figure pulled back from the brink of extinction, a rare environmental triumph rightly heralded by Secretary of the Interior Ken Salazar. "At a time when so many species of wildlife are threatened, we once in a while have an opportunity to celebrate an amazing success story," Salazar said in a press release. After four decades on the endangered species list, he declared, "The brown pelican is back!"

No one's celebrating now. The brown pelican, whose regal mien and fulsome plumage inspired what may be American wildlife painter James Audubon's signature masterpiece, has become tragically emblematic of the BP disaster. What American hasn't been sickened and outraged by the heartbreaking photographs of the Louisiana state bird smothered in oil the nauseating color of sewage, its great wings weighed down with crude?

Hunted for its feathers, devastated by the pesticide DDT and struggling with habitat loss from widespread erosion and subsidence of the Gulf Coast wetlands, the brown pelican was first declared endangered in 1970. After a four-decade fight for survival, the birds became early victims of the BP oil disaster.

Less than two weeks after the blowout, the first oiled pelican was picked up on Stone Island in Louisiana's Breton Sound and flown by helicopter to the bird rehabilitation center set up at nearby Fort Jackson, near the mouth of the Mississippi. Thin and covered by oil over his entire body, the young male pelican was washed in Dawn dishwashing detergent and given intravenous solutions until he was strong enough to eat fish. A week later, he was released into Pelican Island National Wildlife Refuge near Vero Beach, Florida.

Other birds and animals haven't fared so well. By August, more than two thousand oil-soaked pelicans had been picked up dead or dying along Gulf shores. Another 1,200 were found dead without oil on their feathers, though a large percentage undoubtedly had oil in their bodies, after eating fish contaminated by oil.

Biologists use a rule of thumb that at least ten creatures are dead for every one dead animal found. The reason: most simply sink into the ocean. The ocean and wetlands, though, are fiercely competitive and even hostile habitats. That means that when an animal of any sort gets weakened or becomes ill, they become easy prey for predators. We'll never know for certain how many animals died because the oil spill knocked them off their game. We can be sure, though, that the damage already being done to the Gulf from this disaster is as diverse as the range of species there. About sixty dead bottlenose dolphins had been found by the end of August. Most had no oil on their skin, but dolphins must surface to breathe. And the frequency with which pods of dolphins have been spotted swimming in oiled water since the spill suggests that they're breathing toxic fumes at a minimum, and perhaps getting oil down their blow holes and into their lungs.

Oil is the likely culprit in the death of a twenty-five-foot sperm whale recovered in June about a hundred miles from the Macondo site. The sperm whale is one of six Gulf of Mexico whales that are listed as endangered. The others are the blue whale, the humpback whale, the sei whale, the fin whale and the northern right whale. There are so few sperm whales left, their reproductive rates so low and gestation periods so long, that the loss of as few as three females could literally spell doom for these creatures.

The impact on birds will be global. A third of North America's bird species rely on the Mississippi flyway for annual migrations stretching as far north as the Canadian Arctic and as far south as Patagonia. "Tens of millions of shorebirds, waterfowl and other migratory birds will land on oiled beaches, in sullied coastal wetlands and on tainted ocean waters," warns the National Audubon Society.

"Nothing good happens once there's oil in the water . . . From that point on, you're dealing with trade-offs," explained Charlie Henry, an environmental scientist working on the federal response to the Macondo disaster for the National Oceanic and Atmospheric Administration. "All we do every day is to try to minimize the damage."

By the time it was capped in mid-July, the Macondo well had spewed an estimated two hundred million gallons of oil, much of which had spread across a swath of the Gulf the size of South Carolina. Flying out over the ocean in a single engine seaplane a couple weeks before, I gazed down on a chilling site. Running for miles across the sea's surface was a ghoulish line of black and brown, as though someone had dipped a brush in a pot of tar and swiped a streak of toxic sludge across the blue waters below. Behind the line, and stretching as far as the eye could see, the ocean was smothered by a thick, unbroken tarp of oil spreading out from the site of the Macondo blowout, twelve miles to the south.

It wasn't only the browns and the blacks in the ocean that shocked me, the rusts and the rainbow sheen, or even the way the water went dull, like charcoal in places, gun barrel grey. It was the way the water just sat there below us, seeming not to move, like some thick toxic soup gone lifeless and still beneath an unearthly pall of crude.

As we flew on, it got worse, the carpet of oil thickening the nearer we drew to ground zero, the site where the well blew out.

From there, thick surface oil spread out in all directions, as far as the eye could see. By then, oil had washed up along 630 miles of Gulf Coast shoreline across an arc running eastward from near where the Sabine River opens into the Gulf at Louisiana's western boundary with Texas. The oil was in the ocean. It was in the coastal shallows. It was on the beaches, in the wetlands, in the marsh and in the grass. There were tar balls in Lake Pontchartrain.

"This is a catastrophic situation," said Michael Blum, assistant professor of ecology and evolutionary biology with Tulane University in New Orleans. "In terms of scale and intensity, this is a catastrophe." Standing aboard a skiff drifting alongside a marsh near the Louisiana barrier island of Grand Isle, Blum, a specialist in marsh ecology, pointed to a lush field of spartina grass, a brilliant, almost luminescent, shade of green. Where the grass met oil-soaked water, the grass was covered a grimy brown two or three feet high.

"It's essentially killed that first line of plants," said Blum, explaining the vicious cycle of oil reaching into the marsh inch by inch, killing grass with its advance, and the erosion of the fine and fertile silt clinging to the spartina grass, opening the door to accelerating damage from hurricanes and storms.

"This is just going to get progressively worse," he explained. "Then we're talking about potential collapse." At risk, said Blum, is some of the most productive habitat anywhere on Earth. "You're talking about essentially the matrix upon which everything depends," he said. "Everything from basic fisheries—blue crab, brown shrimp, snapper, speckled trout—to the list of critically endangered shorebirds, all make use of this habitat."

Local and regional officials fumed. "We're in a battle,"

Louisiana Governor Bobby Jindal declared. "We're in a war to keep the oil out of our wetlands, off of our coasts," he told reporters in July, after touring oil-soaked delta marsh. "Every day it's out there it's doing damage," he said. "You don't see the bugs. You don't hear the wildlife. You smell the oil."

It took BP eighty-seven days to stop the gusher. In that time, about 4.9 million barrels (two hundred million gallons) of crude oil escaped, according to government estimates.

Stopping the well didn't eliminate the oil. Nearly two weeks after Macondo was capped, the Coast Guard's Thad Allen told reporters there were "hundreds of thousands of patches of oil," still in the Gulf. "When the tide comes in and goes out, you'll have oil into the beaches, in some cases with tar balls or mats," he said. "There's still a lot of oil that's unaccounted for."

Much of that, said Allen, remained out of sight, suspended below the surface. Some was suspended throughout the water column in tiny droplets forming cloud-like pools, or drifting for miles in plumes the size of Manhattan, creating oxygen-depleted, toxic dead zones thousands of feet underwater. The extent of the damage being done by these plumes is unknown, but they would certainly weaken or destroy any of the scores of species of fish that might travel through them for any length of time, as well as deepwater species like sperm whales and squid, and coral reefs and other sea floor habitat that might be enveloped by plumes. As they drift up along the Continental Shelf, they can disrupt highly productive habitat populated by grouper, red snapper and myriad other fish.

Besides the oil, there is the chemical dispersant—1.84 million gallons of which BP had sprayed and pumped into the ocean and at the mouth of the gusher by the end of July. BP applied about two-thirds on the surface and the rest near the head of the well, five thousand feet under the sea.

Dispersants contain solvents that reduce the surface tension of the oil. Dispersants don't change the chemical makeup of the oil, or make it any less toxic. Instead, they break the oil into droplets. One result has been to reduce the amount of oil coming ashore, but to increase the amount that remains suspended in the water, where some has gathered into plumes.

The dispersant itself is toxic. And, sadly, the dispersant being used in the Gulf has advanced little, if at all, in the twenty years since the *Exxon Valdez* disaster, because the oil industry has invested very little since then to improve on the techniques and materials used to clean up a spill. There is no history of dispersant being applied at anything like the amount or the depths that it has been deployed in the Macondo spill, Henry said. In essence, the Gulf has become a giant chemistry experiment, with more questions than answers as far as the use of dispersant goes.

"It is unclear if this will limit the damage from the spill or cause even greater harm," Senator Sheldon Whitehouse (D-R.I.), said at an August 4 hearing of the Environment and Public Works Committee. "We are now seeing large quantities of oil present in the water column, and it could already be starting to settle on the sea floor. We don't know yet what effect this could have on the Gulf ecosystem, from the planktons that form the base of the food chain on up to the apex species, including the bluefin tuna and the sperm whale."

Oil contains toxic chemicals and heavy metals that can harm or destroy wildlife and habitat in multiple ways. Otters, birds and other animals that rely on clean coats of feathers or fur to control body temperature, can suffer hypothermia from being oiled.

Animals can take in the oil by breathing, either through lungs or gills, or ingesting it through food or water.

"It rolls into the estuaries, where everything raises up, your eggs for your crabs, your spat for your oysters, your shrimp," said Ryan Lambert, owner of Cajun Fishing Adventures, a recreational charter boat business in the lower delta town of Buras, Louisiana. "This stuff is killing from the bottom of the food chain up."

In the early months following the Macondo blowout, more than three hundred Gulf area residents sought medical care for maladies ranging from headaches and dizziness to nausea, coughing, respiratory distress and chest pain. Three-fourths of the victims were working on spill clean-up.

While these symptoms can accompany a range of illnesses, they are typical of the kinds of problems resulting from acute exposure to the kind of toxics contained in crude oil and chemical dispersants. Crude oil is a toxic stew of hazardous substances. Benzene, toluene and xylene, for instance, are volatile organic compounds that evaporate and, once in the air, can irritate lungs and the central nervous system. Benzene can cause leukemia in humans, as can naphthalene, another component of crude. Oil can also release hydrogen sulfide gas, another central nervous system threat. Dispersants contain propylene glycol and sulfonic acid salts, which can cause respiratory irritation.

Eleven million gallons of the Macondo oil was torched at sea, the Coast Guard reported, during more than four hundred so-called "controlled burns," during which the oil literally went up in flames from the ocean's surface. That gets oil out of the water, but it puts hazardous chemicals like benzene and hydrogen sulfide into the air, where it can make its way to shore and threaten the health of humans as well as wildlife.

Algae, plankton and tiny crustaceans and fish that come

in contact with oil either perish themselves—denying larger
fish sustenance—or become eaten, the toxics they've ingested
traveling up the food chain. However the oil makes its way into
an animal's system, once there it can cause a wide range of
maladies—brain lesions, internal bleeding, damage to kidneys and
other organs, stress and pneumonia among them—that can lead
to sickness or death.

The Grand Isle mayor is a man of faith, not easily brought low
by hard times. Not when folks sent donations chiseled out of
their Social Security checks—$81 from a retired woman in North
Carolina; $42 from an Ohio couple—to try to help the town
survive. Not when he met with constituents day after day who
couldn't pay their water bill. Not when he looked out at vacant
marina slips during weeks and months they should have been full.
But on the day he saw his coastal waters heavy with oil, something
inside David Camardelle just seemed to give way.

"It brought tears to my eyes," he said, "seeing [globs of]
oil about the size of a pancake coming through, and knowing
there's nothing I can do." Camardelle is the mayor of tiny Grand
Isle, a Louisiana Gulf coast fishing, shrimping and beach town
nicknamed the "Cajun Riviera." In July, he testified before the
panel President Obama appointed to investigate the BP blowout
and advise the administration on next steps.

"It works on you," Camardelle said, describing the anguish,
the anger and the hopelessness he feels when he looks in the
eyes of struggling fishermen, shrimpers, dock workers, restaurant
owners and others he's known for years. "All I tell our people is,
'Look, everyday we'll take it one day at a time.' And, 'stay strong,
pray a lot. God's going to take care of us.'"

Still reeling from the ravages of Hurricane Katrina, which

flattened and then flooded much of the region and killed 1,800 people just five years ago, the Gulf Coast economy is being sacked once again, this time by the catastrophic BP oil spill. Months of commercial and recreational fishing revenues are gone, or very nearly so. Summer beach rental property sits boarded and mute, with oiled waters in many places off limits to bathers and travelers unwilling to chance it. And what economists call "ripple effects" are blasting through the wounded Gulf coast economy.

"This is part of a rolling disaster for the Gulf coast community," said Mayor A.J. Holloway of Biloxi, Mississippi. The entire local fishing industry—from the people who go out on the water to the businesses that supply them with boats and equipment, dock services, ice, tackle and bait—is in "free fall," a University of Southern Mississippi analysis found, with revenues down as much as ninety percent. Along Mississippi's sixty-two miles of beaches, hotel receipts are off fifty percent compared to 2009, and that, said Holloway, "was a terrible year," due to the nation-wide recession. "They were on the verge of coming through Katrina and weathering the recession," he said. "This should have been the breakthrough year. Now they just don't know if they have the financial wherewithal to make it through this."

A July report by Moody's Analytics said the oil disaster could cost the Gulf coast region 17,000 jobs and $1.2 billion in lost revenue just in 2010. Others expect the fallout to be far greater. Oxford Economics, a commercial venture affiliated with Oxford University in the United Kingdom, released a July report concluding that the spill is likely to cost the Gulf coast travel industry alone nearly $23 billion in lost sales over three years. The outfit based the projections on its study of the impact of twenty-five recent disasters, natural and manmade.

Gulf shrimpers have already paid a high price. The oil spill closed most U.S. Gulf shrimping grounds for much of the two-month season for brown shrimp. Typically that provides half the annual income for a Gulf Coast shrimper. Losing that income not only hurts the shrimpers, but others on down the line—dock workers, processors and distributors. "If any part of this link is broken, our industry is going to die," Louisiana Shrimp Association Vice President Acy Cooper told the commission. "After Katrina, we all borrowed money. We all went over our heads to get back in it. This was our year to come back."

The losses will to some extent be offset by clean-up and restoration money coming into the region through BP, state, federal and federal governments and non-profit groups. BP has earmarked $20 billion for Gulf restoration efforts over the next four years, but the total could well exceed that. Many people thrown out of work by the spill quickly found themselves reemployed, making $200 a day or more on BP clean-up crews. Area captains hired out fishing boats to be part of the fleet of four thousand three hundred vessels engaged in skimming oil from the water, the laying down of some 3.5 million feet of containment boom in the water and various other spill-related duties. And while tourism was off, some Gulf coast towns were packed with guests forming part of the 41,200-person workforce put in place to help fight the encroaching oil.

But to Gulf watermen and their families, there's more at stake than a paycheck. What's at risk is a way of life and an essential source of identity for the resilient people of a distinctive region that, in good times and bad, draws its sustenance from the sea. "We are a people in constant disaster," said Derrick Evans. "We don't get to recover from one before another hits."

Sunlight faded over the Biloxi coast, warm winds stirred the white sands. And Evans gathered around a conference table with two dozen church and community leaders to discuss the impact of the spill over brownies and sweet tea. "We don't have a lot of money," said Evans, in a voice that rumbled a bit like thunder off the bayou. "The capital that we do have, and use to the best of our ability, is our relationships and our stories. We didn't know the next chapter was going to include this. It's almost like a book of Job that doesn't end."

Chapter 3

Big Oil, "Small People"

On June 16, with BP's runaway Macondo well gushing unabated into the Gulf of Mexico and a region of fourteen million people cast into turmoil, President Obama summoned the oil company's chairman, Carl Henric Svanberg, to the White House.

The Gallup poll showed Obama's job approval rating slipping to just forty-five percent, and the president wanted to send a message to the public—and directly to BP's top brass—that he was directly engaged on the blowout. Svanberg needed to shore up investor confidence in a company whose stock price had plunged by half, washing out a breathtaking $90 billion in market capital in just six weeks, amid a barrage of relentless and increasingly rancorous criticism from Congress. Both men hoped to turn things around.

Obama got what he wanted. BP pledged to put up a $20 billion contingency fund to cover damages to the Gulf and its people and agreed to let it be managed by Kenneth Feinberg, the mediation attorney who oversaw the compensation fund for victims of the attacks of September 11, 2001.

Svanberg didn't do quite so well. Standing in front of the West Wing afterwards, before a gaggle of reporters, he was asked how long he met with the president. "I spent a fair amount of time," Svanberg replied without elaboration. Then he pivoted to take advantage of the cameras for a little public relations work. "He is frustrated, because he cares about the small people," Svanberg said of Obama. "And we care about the small people. I hear comments sometimes that large oil companies are greedy companies, or don't care, but that is not the case in BP. We care about the small people." The reaction on the ground was immediate.

"On this side of the pond, everybody's an equal partner, and there are no small people or big people," New Orleans Mayor Mitch Landrieu shot back in a press conference the next day. Beside him stood a French Quarter oyster shucker, the head bellman from the Royal Sonesta hotel and a tuba player from the Preservation Hall Jazz Band. "We are American citizens," said Landrieu. "There are no small people in this epic battle."

Svanberg apologized, saying he'd spoken "clumsily." And a BP spokesman shrugged off the slight as a casualty of translation: Svanberg, who is originally from Sweden, simply tripped up on his English. Svanberg is the chairman of the board of the largest corporation in Britain and the fourth-largest in the world. It is a safe bet he pretty much means what he says, not unlike Tony Hayward, who has a gift for the telling quote that suggests where the oil giant's highest priorities lie.

Warning from the start that Americans would try to exploit the disaster in the Gulf, Hayward told *The Times* of London in early May "I could give you lots of examples. This is America— come on. We're going to have lots of illegitimate claims. We all know that." Three weeks after the blowout, he observed to the

Guardian newspaper that "the Gulf of Mexico is a very big ocean."
In an interview broadcast the next week on Sky News, Hayward
said that "the overall environmental impact will be very, very
modest." On Sunday, May 30, Hayward diminished the health
concerns of hospitalized clean-up workers and denigrated the
findings of accomplished oceanographers.

After nine cleanup workers were hospitalized complaining
of respiratory ailments, nosebleed and headaches, Hayward told
CNN it could be "food poisoning" that struck, and not the toxic
fumes of oil and chemical dispersant they'd been exposed to on
the Gulf. He disputed the findings, later confirmed by the National
Oceanic and Atmospheric Administration, of marine researchers
from the University of South Florida and the University of
Georgia, who found plumes of oil up to ten miles long and three
miles wide in Gulf waters up to three thousand feet deep. "The
oil is on the surface," Hayward told the Associated Press. "There
aren't any plumes." On the same day, in an interview broadcast
on Fox News, among other outlets, Hayward said this: "There's
no one who wants this thing over more than I do. You know, I'd
like my life back." Hayward later apologized, calling the remark
"hurtful and thoughtless."

If oil executives like Svanberg and Hayward think they're special,
perhaps it is because that is how we treat them. We have asked
the companies they run, after all, to provide us with the one
commodity we keep telling them we cannot live without, no
matter the cost or the risk. We say it each time we fill up our gas
tank, each time we get on a plane, each time we buy food, clothing
or anything else that's made and shipped cheaply to our homes
from around the country, or around the world. Our demand has
enabled the oil companies to build an industry that is profitable,

powerful and influential in ways we seldom understand. Our entire civilization is so heavily dependent upon the companies that provide us with oil, that we seldom even question what they do or how they do it. In some quarters it is almost as if oil itself has become something of a modern god.

"If oil didn't exist, we'd have to invent it," said Robert Bryce, senior fellow with the Manhattan Institute for Policy Studies, an economic think tank based in New York. "There's nothing else that comes close in terms of energy density, ease of use, transportability. It's really a miraculous substance."

Oil, Bryce explained, contains a lot of energy in a concentrated form that weighs relatively little and is fairly portable. That is why it is used in our cars, airplanes, trucks and ships. By some measures, a forty-two-gallon barrel of oil contains the energy equivalent of eleven thousand man-hours of labor. That means a twenty-gallon tank of gas delivers the energy equivalent of a single laborer working for five years. Oil is also the feedstock for the plastics we use for everything from picnic forks to computer keyboards, as well as synthetic fabrics ranging from nylon to Polartec and a wide array of chemicals essential to the foods, medicines and industrial processes we rely on.

With less than five percent of the world's population, the United States is by far the world's largest consumer of oil. Today, tomorrow and every day, Americans will burn eight hundred million gallons of oil—one out of every four gallons produced everywhere in the world. That is as much as will be used in the four next-largest oil consuming countries combined. And those countries—China, Japan, India and Germany—make up forty percent of the world's population. In the United States, we will burn twelve percent of the world's daily oil output just to drive our cars and trucks. Today, tomorrow and every day.

"We have a serious problem," President George W. Bush, an oilman and former governor of Texas, told the nation in his 2006 State of the Union address. "America," he said, "is addicted to oil."

Rather than working to reduce our dependence on oil, we've asked the oil companies to feed it, with little concept of what that requires—from us or the people who provide us with oil—or what it demands in return. Rather than using that oil carefully and efficiently, in ways that reflect both the remarkable substance it genuinely is and the challenges and risks in obtaining it, we often waste it in inefficient gas-guzzling cars and trucks and other energy drains.

"Most Americans have no idea, nor do I think anybody involved in any one part of the industry has any idea, of the scope of the politics that's required, the technology, the sheer operations, the upkeep, the livelihoods of the workers, the environmental protections and harm," said Antonia Juhasz, author of the 2008 book *The Tyranny of Oil.* Said Juhasz, "It's such a big, broad industry, such a multi-tentacled industry, that it's kind of overwhelming."

In 2010, Americans will spend $850 billion on petroleum products, assuming an average retail price of $2.90 per gallon for gasoline and the other fuels that account for about ninety percent of the oil we use. That's about four times as much as we will pay every policeman and public schoolteacher in America, combined. We could take the entire annual paycheck of every teacher at the blackboard, and every cop on the beat, everywhere in America, and it would buy us, at retail prices, just three months worth of oil.

Because half our oil is imported, it holds our entire economy hostage to even relatively minor shocks of global price and supply.

It also accounts for about a third of our trade imbalance with the rest of the world. That gap comes at the price of American jobs. There isn't a direct dollar-for-dollar correlation. But when there's a trade imbalance, we have to borrow money to make up the difference—money that would otherwise be available for investment or production here at home. Between 2005 and 2009, the United States imported $3.4 trillion more worth of goods than it exported. Oil imports—mostly crude oil—made up $1 trillion of that deficit, after accounting for the petroleum the country exports.

Imagine how many American jobs might have been created if that money had stayed in this country. Imagine how U.S. competitiveness might have increased if that money had been invested here. Imagine how much better equipped we might be to compete in the global marketplace. Over time, the impact of oil imports has been even greater. In the sixteen years between 1994 and 2009, Americans bought $7.7 trillion more goods from abroad than we shipped overseas. Oil accounted for $3 trillion of that, making up forty percent of the U.S. trade gap.

Our dependence on oil also distorts U.S. diplomacy. It puts us in league with countries like Saudi Arabia, Venezuela, Russia and other oil exporters that don't always share our values or goals. It puts us in direct competition with nations like China and India, where economic growth drives soaring oil consumption. And it imbues us, as a people, with the odor of complicity in the kinds of bloodshed, human rights abuse and economic deprivation endemic to places like Nigeria, Sudan and Myanmar, where oil revenues have sustained corrupt and inept leaders for decades.

There may be no one who understands this better than former U.S. Secretary of State Condoleezza Rice. Before President George W. Bush named her the nation's top diplomat, she served as his National Security Advisor. She served on the staff of his father's National Security Council before that and, in between those jobs, she was a member of Chevron's board of directors.

"Let me be very clear about the search for oil," Rice said in a speech before the Chicaco Council on Foreign Relations in April 2006. "It is distorting international politics in a very major way. It's distorting because there are places that have oil that are using oil as a weapon, or using oil as a carrot for certain policies, and that's troubling."

Securing global oil supplies and transportation routes adds to the already onerous burden borne by the men and women who defend American interests abroad, often against adversaries funded by petrodollars. And it is driving the companies and the people we buy our oil from to push the limits of what can be done safely, to venture into increasingly difficult environments, like the deepest waters of the Gulf of Mexico, and to put irreplaceable habitat increasingly at risk.

"It's not understood at all by the vast majority of the public. All the public wants is cheap, always-available motor fuel, and they don't ask how it gets there," said the Manhattan Institute's Bryce, who is also managing editor of *Energy Tribune*, an online news magazine. "What BP was doing at the Macondo well is the marine equivalent of the space program," said Bryce. "You are asking a tremendous amount of them."

The oil companies, in turn, ask back. In 2009, the four largest U.S. oil companies—Exxon, Chevron, ConocoPhillips and Valero—had combined sales of $658 billion. Add the Dutch oil giant, Shell, and Britain's BP, and the total comes to $1.2 trillion. Those are global sales, but, to put the figure in perspective, it's the equivalent of 8.4 cents for every dollar spent in the United States that year. If those six oil companies were a country, that country would tie Russia for the world's twelfth-largest economy, just ahead of Australia.

The year 2009, moreover, was an off-year for oil. Revenues were down dramatically due to recession and falling crude oil prices. In 2008, those six companies had combined sales of $1.9 trillion. Only seven countries in the world have economies larger than that. That kind of throw weight, the industry claims, is part of what keeps the U.S. tank topped off. "The oil and natural gas industry is massive," states the American Petroleum Institute, the industry trade association, "because it has to be to effectively compete for global energy resources."

At the same time, when the stakes for a single industry are that high, its influence looms large. Since 1998, the oil and gas industry has spent $1 billion to lobby the U.S. Congress and the administration in Washington, including a record $168 million in 2009 alone, according to the Center for Responsive Politics, an independent organization that tracks such outlays.

In addition, individuals and political action committees affiliated with oil and gas companies have donated $238.7 million to political parties and candidates since 1990. Most of that— seventy-five percent—has gone to Republicans, though Obama received $884,000 from oil and gas interests during his 2008 campaign, the Center for Responsive Politics has found.

That doesn't include hundreds of millions of dollars the industry has spent lobbying officials in Texas, Louisiana, Alaska and other states. Nor does it cover advertising intended to influence public opinion on topics such as global climate change, offshore drilling, tax policy and fuel efficiency standards for automobiles.

The Washington oil lobby isn't just a few political roustabouts. It's a seasoned and well-heeled army of nearly seven hundred Washington insiders that epitomizes the revolving door approach to influence peddling. Three out of every four lobbyists registered to represent oil and gas interests once worked in Congress or for a presidential administration, a July analysis by *The Washington Post* disclosed. Their ranks include former House veteran Bill Archer, the Texas Republican who chaired the powerful Ways and Means Committee, where all tax provisions originate, as well as former U.S. Senator John Breaux, the Louisiana Democrat who served on the Senate Finance Committee. Among those registered to lobby for BP: Ken Duberstein, who was chief of staff for former President George H. W. Bush. And of course some politicians have their own background in the oil industry before coming to Washington. President George H. W. Bush and his son, George W. Bush, were both executives in Texas oil operations before going into politics, as was former Vice President Dick Cheney.

"The oil industry is omnipresent in Washington . . . always providing information, always hinting at post-government-service jobs, and always applying pressure," wrote California's former Environmental Secretary, Terry Tamminen, in his 2009 book *Lives Per Gallon: The True Cost of Our Oil Addiction.*

The industry gets its money's worth and then some. Between 2002 and 2008, the oil and gas industries received $51 billion in federal subsidies and favorable tax treatment, according to a 2009 study by the Environmental Law Institute, an independent Washington research organization. Those subsidies are above and beyond the traditional investment and manufacturing tax breaks available to businesses in general, including those in the oil industry.

"It's been going on for almost a hundred years," said John Pendergrass, senior attorney with the institute and a co-author of the 2009 study. Part of the subsidies are embedded in the tax code, meaning they are not subject to the Congressional scrutiny that takes place each year during the budget debate. For example, a 2005 study by the Congressional Budget Office found that oil and gas companies pay a 9.2 percent tax rate on wells and other structures, about a third of what most businesses pay on their assets.

That kind of favorable treatment will cost taxpayers $20.4 billion between 2010 and 2019, the U.S. Treasury Department estimates. Obama has asked Congress to phase out the subsidies, a proposal the industry is lobbying hard to defeat. The American Petroleum Institute argues that the tax breaks encourage companies to take financial risk, find oil and employ workers. The Obama administration counters that oil profits justify the risks of searching for oil and the industry employs far fewer people per dollar invested than corporate America at large.

Together, Shell, Exxon, BP, Chevron, ConocoPhillips and Valero employ 400,000 people in more than one hundred countries around the world. The oil industry, though, is not a labor-intensive business. It's ten times more capital-intensive, in fact, than U.S. industry at large, according to the Treasury Department. Oil companies rely on harnessing sophisticated equipment to valuable resources, and managing the investment required to do that. Together, Shell, Exxon, BP, Chevron, ConocoPhillips and Valero own $1.1 trillion in assets. That's just fifteen percent shy of the total value of a year's economic output in India, the second-most populous country on Earth. The idea that an industry that well endowed requires favorable tax treatment is something the Obama administration rejects.

"The tax subsidies that are currently provided to the oil and gas industry lead to inefficiency by encouraging an over-investment of domestic resources in this industry," Alan Krueger, the Treasury Department's assistant secretary for economic policy and chief economist, testified in September, 2009, before the Senate Subcommittee on Energy, Natural Resources and Infrastructure.

Under the U.S. tax code, in fact, BP could end up saving $10 billion in oil spill costs. On July 27, BP told shareholders it would write off $32.2 billion in losses related to the Macondo blowout. That includes $2.9 billion for BP's response as of that date and $29.3 billion in expenses it expects in the future, including the $20-billion escrow account. The write-off could save BP about $10 billion in U.S. taxes. In effect, American taxpayers would be footing the bill for half the BP escrow account.

"I was appalled," Senator Bill Nelson, a Florida Democrat, wrote two days later in a letter asking Senate Finance Committee Chairman Max Baucus to investigate. "As a basic policy matter, is it appropriate for BP to claim a deferred tax benefit as a result of any negligent actions in the Gulf of Mexico?"

Tax treatment isn't the only, or even the most costly, public benefit enjoyed by the oil companies. For one thing, many of the public health and environmental costs of producing and burning oil are not paid by oil companies or consumers. Toxic waste, emissions from refineries and exhaust from cars and trucks have been shown to cause, or exacerbate, a long list of ailments—cancer, asthma, respiratory and cardiovascular disease among them—in ways that impose real social and economic costs.

Chevron is the target of the largest environmental lawsuit in history, with up to $27 billion at stake over damage done to pristine Amazon jungle when Texaco—since bought by Chevron—produced oil in Ecuador. The government in Quito signed off on Texaco's $40-million cleanup program after the company pulled out in 1992. The area, though, remains ravaged to this day, with open pools of crude oil, drilling mud and other toxic remnants of the industrialization of the jungle. Indigenous people have filed suit. Chevron has launched counter charges of government corruption and fraud, saying the case has "descended into a judicial farce." The company blames the enduring squalor on the government of Equador, and its state oil company, Petroecuador, charging that they've failed to do their part in the cleanup. The case is pending before a small court in Ecuador, whose president, Rafael Correa, has called the pollution "a crime against humanity."

When the price of a product such as oil doesn't fully reflect the costs it imposes on the environment or society at large, economists say those costs have been "externalized." That means those costs are being paid by people who don't benefit from the product. In practice, they may not be bearing the costs willingly; they may not even be aware of the costs they're bearing. It's not a subsidy, strictly speaking. If those costs were reflected in the true price of oil, though, we'd all be burning much less of it.

"Oil and natural gas prices, for example, do not reflect the environmental harm caused by the release of greenhouse gases in the atmosphere associated with oil and gas production and consumption," Krueger said in his senate testimony. "In addition, the price of oil does not reflect the risks associated with U.S. oil dependency or the costs of traffic congestion."

What Krueger called "the risks associated with U.S. oil dependency" go well beyond economic. Short of food, water and physical security, there may be no more strategic commodity for the United States than oil. The quest for stable supplies has helped shape diplomatic and national security policies of the United

States—and, indeed, every major world power—for most of the past century.

U.S. dependence on imported oil has long been recognized as the Achilles heel of American security. It increases "strategic vulnerability" and constrains the nation's ability to pursue "A broad range of foreign policy and national security objectives," a task force of the Council on Foreign Relations concluded in a 2006 report, "National Security Consequences of U.S. Oil Dependency." "The longer the delay" of freeing ourselves from oil dependency, "the greater will be the subsequent trauma," the task force warned.

American defense policy, in some ways, begins with oil. The U.S. military is the single largest consumer of oil in the world, burning 16.8 million gallons of it every day—at an annual cost of $11.5 billion—to fuel fighter aircraft, battle tanks, warships and other machinery, and to operate more than 300,000 buildings in the United States and around the world. The Defense Department, in fact, uses more oil than all but thirty-four countries, according to Central Intelligence Agency rankings. Its largest supplier: BP, which last year sold the Pentagon $2.2 billion worth of fuel, twelve percent of the military's total consumption.

The military's own oil dependency can be a battlefield liability. A Humvee gets four miles to the gallon. An Abrams battle tank burns five gallons of gas every three miles. Keeping the equipment fueled is dangerous work in a war zone. In 2007, 170 American troops were killed in fuel convoys in Iraq, according to Congresswoman Gabrielle Giffords (D-Ariz.), who introduced a bill in May aimed at reducing the Pentagon's reliance on oil.

It's not known how much of U.S. defense spending—$655 billion this year alone—is attributable to safeguarding American oil supplies in volatile regions like the Middle East or ensuring safe passage through sea lanes and critical shipping choke points like the Strait of Malacca, the Suez Canal or the Gulf of Hormuz. When Iraq invaded neighboring Kuwait, however, on August 2, 1990, in response to a long-unresolved dispute about control of vast oil reserves near the two countries' border, the United States was quick to respond. With 4.8 trillion gallons of crude oil, Iraq has eight percent of the world's proven reserves. Kuwait has seven percent. Had Iraq prevailed, it would have controlled fifteen percent of the world's known crude oil reserves. More than that, Iraq had hundreds of tanks and 400,000 troops moving south toward Saudi Arabia, posing a direct threat to the largest oil-producing nation in the world, a country that sits above eighteen percent of the world's proven crude reserves.

"This will not stand, this aggression," President George H. W. Bush declared to reporters at the White House on August 5, 1990. Two days later, two squadrons of U.S. Air Force F-15 fighter jets arrived in Saudi Arabia, as two U.S. aircraft carrier battle groups arrived in the region. By December, there were seven hundred U.S. warplanes and 230,000 American troops in the area, part of a massive buildup of U.S. and allied military forces in preparation for all-out war that evicted Iraqi forces from Kuwait the next year. In National Security Directive 45, dated August 20, 1990, the Bush administration laid out the case for war.

"U.S. interests in the Persian Gulf are vital to the national security," the directive read. "These interests include access to oil and the security and stability of key friendly states in the region." Those objectives were reaffirmed in the Pentagon's latest mission statement, the Quadrennial Defense Review, released in February 2010.

The document listed "energy security initiatives" as an overarching military responsibility, along with global peacekeeping, stability and reconstruction operations, missile defense cooperation and nonproliferation efforts. It specifically mentioned the oil-rich Middle East, noting that stability there "remains critical to U.S. interests." And it cited the need to protect "vital sea lines of communication" in the Indian Ocean, the sea path that links the United States and its European and East Asian allies to half of the world's supply of oil.

U.S. forces, of course, returned to Iraq in March 2003, initiating a war that persists to this writing. Before launching the U.S.-led invasion, President George W. Bush spent months telling the American people, and U.S. allies around the world, that his goal was to rid Iraq of chemical and biological weapons. In the wake of the attacks of September 11, 2001, which killed nearly three thousand in New York, Washington and Pennsylvania, Bush said the country could not run the risk that weapons of mass destruction might fall into terrorist hands. His administration rejected speculation from critics who alleged that a separate goal of the invasion was to secure U.S. access to oil in Iraq.

After more than three years of fighting, though, Bush appealed to the American public for continued support of a grinding war that eventually would take the lives of more than four thousand four hundred American troops and leave thirty-two thousand wounded. In formal remarks opening a news conference in the East Room of the White House on October 25, 2006, Bush introduced a new rationale for prevailing in Iraq. In addition to keeping weapons of mass destruction out of the hands of al Qaida and its ilk, it was vital to deny them Iraq's oil.

"If we do not defeat the terrorists or extremists in Iraq, they will gain access to vast oil reserves and use Iraq as a base to overthrow moderate governments across the broader Middle East," Bush said, later warning of the dangers of "a world in which oil reserves are controlled by radicals in order to extract blackmail from the West." Iraq's store of oil is vast indeed; it has 4.8 trillion gallons of proven reserves. Only three countries in the world have more: Saudi Arabia, Canada and Iran.

In the first four months of this year, the United States imported about twenty million gallons of oil a day from Iraq, 2.5 percent of U.S. demand. Some of it comes from BP, which is producing forty-two million gallons a day from an Iraqi oilfield it shares with its partner, the China National Petroleum Corporation.

After decades on the sidelines of the global economy, China, the most populous nation on Earth, has had annual economic growth in the range of ten percent for most of the past quarter century. In 1993, China began burning more oil than the country could produce, importing the difference. Since then, its consumption has risen to about 336 million gallons a day, about half of which comes from exporters like Saudi Arabia, Angola, Iran and Russia.

Over the next twenty-five years, China's oil consumption will nearly double, as will happen in Asia's other developing giant, India. That's the main reason why global oil consumption is on track to hit a staggering 4.7 billion gallons a day by 2035, according to the U.S. Energy Information Administration. That means that, unless we begin to use this precious oil more wisely and carefully, someone will have to find and produce an additional 1.6 billion gallons a day of oil, just to meet projected demand.

Between now and 2020, though, the industry estimates enough existing oil wells are going to run dry to take 1.7 billion gallons a day off the table. That will have to be made up just to keep the global fuel tank from running on empty. Not only must the oil industry increase its output by about fifty percent by 2035, it must do so while finding new wells to replace those producing today.

"Most of it," Shell's chief executive officer, Peter Voser, told the London Business School in July, "will need to come from resources that haven't even been found yet." The search, though, is remaking the map of the world.

Over the past three decades, global climate change has melted enough Arctic sea ice to cover the entire United States east of the Mississippi River. In the summer of 2007, Russia decided to take advantage of the opening seas, sending a submarine to plant a Russian flag on the seabed of the North Pole. Critics say this was a land grab in one of the last great oil frontiers.

"With the melting of the Arctic, now huge oil resources, perhaps larger than those of Saudi Arabia, have been uncovered," Indiana's Richard Lugar, the ranking Republican on the Senate Foreign Relations Committee, said in February. "The Russians immediately sent a ship up and planted a flag. This was a grandiose gesture. It doesn't establish that they're going to be drilling shortly. But the fact is, this is going to be an area of huge contention," Lugar said at a committee hearing. "We've got to pin down our sovereignty."

As global oil consumption continues to rise, such friction seems bound to increase. Either we will find a way together to grow our economies without using so much oil, or we will be forced to compete with increasing intensity for a strategic resource growing ever more scarce. Cold war, fractious disputes and armed conflict will break out over energy rivalries, with no country more vulnerable than our own.

"The underlying problem is the high and growing demand

for oil worldwide," the Council on Foreign Relations task force concluded. "At best, these trends will challenge U.S. foreign policy; at worst, they will seriously strain relations between the United States" and its oil-guzzling rivals.

"These hard truths, the realities of the new energy future, are stark," said Shell's CEO Voser, with oil "steadily becoming harder and more expensive to access and recover." What that looks like has already been previewed in the Gulf of Mexico. There, beneath deep blue waters, our thirst for oil has been joined to an industry that sometimes describes the world as a place where oceans are too big to worry about and people too small to matter.

Chapter 4

Blind Faith

A decade ago, with world oil prices hovering around $26 a barrel and demand steadily on the rise, BP launched one of the most ambitious drilling ventures in history. The gambit was so complex and fraught with potential peril that the company brought in a competitor, Chevron, to share the risk. In June, 2001, the partners struck oil beneath more than five miles of ocean, salt canopy and stone 120 miles off the coast of Louisiana. It was, at the time, the deepest well in the Gulf of Mexico, an endeavor that tested and pushed the technological and operational limits of the oil industry and, indeed, the modern world. They named the project Blind Faith.

"To meet the future energy demands it's very important that we have access to the deeper-water areas of the Gulf of Mexico, such as Blind Faith," Chevron's Gordon Rorrison, the project's engineering manager, explained at the time. "It was a very challenging project, and also very, very rewarding."

It would be hard to find a more fitting symbol of the oil industry's steady and assertive advance into the Gulf's deep waters, or the corporate thinking behind it. As demand for oil has increased, and competition for new sources grown keener, the search has intensified, as has the risk. At precisely the same time, though, government oversight has weakened, the product of lax enforcement of basic safeguards and decades of political push back against common sense protections once deemed essential. We assumed, as a nation, it would all work out. We were acting on blind faith.

"Where I was wrong was in my belief that the oil companies had their act together when it came to worse-case scenarios," President Obama said in press conference a month after the Deepwater Horizon disaster. "Those assumptions proved to be incorrect."

In 1989, the year President Ronald Reagan left the White House, deepwater wells—those drilled beneath one thousand feet or more of water—produced just four percent of all Gulf oil. That figure rose to forty-five percent over the next decade and sixty-five percent in 2004. By the time the Macondo well blew out, eighty percent of the oil produced in the Gulf of Mexico was coming from deepwater wells, including some drilled in water nearly two miles deep.

Industry officials have been clear about the increasing risks that come with building complex systems on the sea floor. Robotic subs work in dark water at temperatures near freezing under pressures twice as powerful as the machinery used to flatten junked cars. Hot oil roils out at up to 275 degrees Fahrenheit. "The deepwater environment is unforgiving, the challenges are immense and the deeper you go, the more difficult it gets," Shell's director for projects and technology Matthias Bichsel said in a December 2009 speech at Qatar University.

And yet, just as oil companies were probing ever deeper

into the Gulf, political support largely broke down for the kinds of essential protections only our national government can provide to defend our safety, environment and health. In one area after another—from the hands-off approach to Wall Street to the fundamental weakening of bedrock safeguards for our air, water, wildlife and lands—government oversight was diminished and dimmed, often for the benefit of the very institutions whose excesses must be vigilantly guarded against. The situation "fostered a climate of complacency," said Robert Percival, director of environmental law at the University of Maryland School of Law. "You had very little government oversight of what the oil companies were doing in the Gulf [and] . . . the oil industry did not feel it was necessary to evaluate the environmental impact of its increasingly risky activities."

Those trends converged with a vengeance at BP's Macondo well. No less than the geologic pressures, corporate judgments and operational shortcomings that contributed to the blowout, it was the push to find oil at any cost, and the absence of effective oversight to keep the threat of disaster in check, that set the stage for catastrophe.

"It was a combination of anti-government sentiment and public and corporate demand for both profits and more oil," explained Larry Sabato, director of the Center for Politics at the University of Virginia. "You put it all together and you create the climate that can lead to an under-regulated system that allows for a disaster," Sabato said in a telephone interview. "The BP spill is a big price to pay."

On January 10, 1901, near the east Texas town of Beaumont, wildcatters uncapped a historic gusher at a place called Spindletop. It blew out thirty-four million gallons of crude oil in eight days, ushering in what some call the "Century of Oil" and bringing unfathomable wealth to a handful of backers. One of them was an industrious Pennsylvania oil and coal magnate, Colonel James M. Guffey.

Six years later, one of Guffey's employees noticed methane—natural gas —bubbling up through the coffee-colored shallows of Caddo Lake in northeast Louisiana. Guffey's company drove cypress pilings into the bottom of the lake, built a rough-hewn derrick on top, and in 1911 hit oil in what the state of Louisiana purports to be the first well drilled from an overwater platform anywhere in the world. Guffey named his company Gulf Oil, later bought by Chevron.

By the 1920s, Louisiana's swamps, bayous and shallow lakes were dotted with similar platforms and rigs, as the state became ground zero for the budding offshore oil industry. In 1937, two companies teamed up to wade into bigger waters. Driving piles of yellow pine into water fourteen feet deep, the Pure and Superior oil companies struck crude a mile off the Louisiana coast. Over the next thirty years, the so-called "Creole Field" survived hurricanes and even war to produce 170 million gallons of oil, heralding the era of a new industry that has since drilled fifty thousand offshore wells in the Gulf of Mexico. By the time Dwight Eisenhower became president in 1953, there were dozens of oil production platforms in Gulf waters up to seventy feet deep, and the long march to deeper waters was on.

"The push out to produce oil in the deeper waters of the Gulf reached the one hundred-foot mark in 1957 and then quickly moved on out to 225 feet in 1965 and more than three hundred feet in 1969," wrote the authors of a Louisiana State University history of the offshore oil industry. By 1970, there were more than a thousand offshore platforms in the Gulf.

In venturing into the Gulf, the oil industry developed technology and techniques not only for drilling beneath the ocean, but also for building the structures and laying the pipeline needed to support oil development and production. Drilling rigs and production platforms had to withstand harsh conditions around the clock and the regular forces of tropical storms, heavy seas and hurricanes. The engineers and scientists who developed these technologies were innovators and risk-takers who drew confidence, not from some government agency providing oversight and guidance, but from their own growing ability to work through problems and overcome great odds. When adversity struck, the oil companies looked to each other, not to Washington, for the solutions required to get at the oil.

"When problems arose, engineers and construction specialists within the individual oil companies joined forces with their counterparts in offshore construction and service companies to solve them 'on the run,'" the authors of the LSU overview wrote. "This 'entrepreneurial' approach was possible in a largely unregulated environment in which the companies enjoyed great freedom to make their own choices."

In 1975, Shell became the first company to find oil beneath water more than one thousand feet deep, at a site it called Cognac, twenty-eight miles off the mouth of the Mississippi. Five years later, Exxon became the second, going five hundred feet deeper than that. Over the next two decades, the industry drilled scores of deepwater wells into more than three dozen oil fields, to depths as great as a mile. And in 2000, for the first time, deepwater wells delivered more oil than those in shallower waters.

The Gulf produces roughly seventy million gallons of crude oil each day, about one-third of total U.S. oil production. Another ten million gallons per day of oil is projected to come on stream over the next three years. Practically all the new oil will come from deepwater wells, which will be supplying as much as ninety percent of Gulf of Mexico oil by then, according to U.S. government projections.

"What we're seeing here is the start of a new frontier in the Gulf of Mexico," Shell team leader Bill Townsley told the *Houston Chronicle* last spring, while standing atop the company's Perdido oil and drilling platform. Two hundred miles off the coast of Texas, Perdido floats in water nearly two miles deep, high overtop one of the largest pools of oil anywhere in the world.

Perdido—the word means "lost" in Spanish—is a kind of industrial monument at sea, fifty thousand tons of technological testimony to the future of offshore oil. The deepest ocean production platform in the world, Perdido is a $3 billion partnership among Shell, BP and Chevron. The platform itself is nearly as tall as the Empire State Building. It bobs like a giant bottle that can take in oil from thirty-five separate wells tapping three different oil fields spread out beneath twenty-seven square miles of ocean floor.

In industry parlance, Perdido sits atop an "elephant," a super-sized field of crude oil, natural gas and other energy-rich hydrocarbons stored in rock formations laid down sixty-five million years ago. Nobody knows for sure how much is there, but Shell estimates it's the hydrocarbon equivalent of at least 126 billion gallons of oil, and perhaps as much as six hundred billion gallons.

"Why are they out there? They're there because that is where the resource is," said Robert Bryce of the Manhattan Institute. "For any of these huge companies to move the needle in any meaningful way, in terms of the resource base, they have to be looking for elephants."

And the elephants are in the deep water. "That part of the basin is older, and the sediments are older and thicker, the packing of sediments containing the hydrocarbons is thicker," explained Jeff Williams, scientist emeritus with the U.S. Geological Survey. "The deeper parts have been accumulating sediment for probably fifty or sixty million years."

That history left behind giant reservoirs of oil and gas in ancient sands miles below the sea. The scramble is on to find them.

Over the past ten years, Exxon has drilled thirty-five deepwater wells in the Gulf, twenty of them in water more than half a mile deep. In March, Shell found a huge pool of oil in water a mile and a half deep on a site it calls Appomattox. It was the company's fourth major deepwater Gulf discovery in less than two years, including one with the equivalent of more than 8.4 billion gallons of oil. Chevron is working a field more than twice that large at a site it calls Tahiti, 150 miles offshore in sands folded beneath a blanket of Jurassic salt up to three miles thick.

And then there's the largest Gulf producer of all, BP, which produces seventeen million gallons of oil a day there, a fourth of the Gulf's total output and ten percent of BP's production worldwide, according to the company's 2009 annual report. BP believed in the deepwater Gulf at a time when some oil companies didn't, a time when the game was beginning its accelerating shift into deeper and more treacherous waters.

"When BP went into the deepwater Gulf in the early 1990s, the area was known within the industry as 'the dead sea,'" BP's chief executive at the time, Tony Hayward, said a month before the blowout. In a speech at the Peterson Institute for International Economics, a Washington think tank, he said BP has stayed the course, developing the technology needed to operate in the Gulf. "It led to a series of extraordinary discoveries."

For an oil company to survive, it must find more oil each year than it sells. BP has managed to do that for the past seventeen years in a row, and the Gulf of Mexico is a big reason why. In September of 2009, BP hit an elephant well in deep water 250 miles southeast of Houston, on a site it named Tiber. A company press release called the reservoir a "giant," and officials said it could be even larger than BP's nearby Kaskida field, which holds more than two hundred billion gallons of oil equivalent—a mixture of oil, gas and other hydrocarbons, expressed, for consistency's sake, in the energy value of oil. The company and its investors had similar hopes for Macondo.

"If the Macondo drilling project had been successful, the oil well could have become the most productive in the region," the Zurich-based investment firm Credit Suisse speculated in a June 15 analysis. "Good times and bad times sometimes come hand in hand."

So it would seem in the deepwater Gulf. What draws adventurers ever deeper, in fact, is the promise of one day tapping into the kind of field BP uncorked a decade ago, at a site the company named Thunder Horse. With a production platform the size of the U.S.S. *Enterprise* tapping into more than eight separate wells, the field pumps out 12.6 million gallons of crude oil a day. At $75 a barrel, that's nearly $30 billion worth of oil a month.

Thunder Horse, though, is also the kind of operation that keeps oil executives up at night. It narrowly averted a Macondo-style blowout in 2003, when a development well was being drilled there. During those operations, the riser pipe connecting the well to the drilling rig snapped. The blowout preventer saved the day, shearing the pipe and shutting down the well.

"No one was hurt, and the well was secure, but the initial scene was daunting," a BP exploration and production advisor and a response and restoration official with the National Oceanic and Atmospheric Administration wrote in a report on the incident. "Loss of containment" the authors wrote, would have been disastrous, resulting in more oil spilled in a week than was lost during the *Exxon Valdez* spill.

In July 2005, three months after being towed into place on its maiden voyage, the Thunder Horse platform was badly damaged after Hurricane Dennis roared through with one hundred mile-per-hour winds. The platform had been evacuated and faulty equipment caused flooding in its massive support pontoons. Fifteen stories high, Thunder Horse listed over as if it might sink. Repairs took nearly a year. By then, tests showed leaks in pipes and massive manifolds along the seabed. Thousands of tons of equipment had to be hauled up, fixed, and then reinstalled using robots roving the ocean's floor. The repair itself was a near-epic undertaking that required the company to develop special techniques and tools. It was June 2008 before Thunder Horse produced the first gallon of oil.

Critics said the problems at Thunder Horse were indicative of a company that was pushing too hard too fast in the deep waters of the Gulf, even as BP was cutting staff and slashing costs in the wake of expensive takeovers of U.S. competitors Amoco and Atlantic Richfield.

There were even greater causes for concern. In March 2005, as Thunder Horse was being readied for service, BP's largest refinery, a 19.3-million-gallon-a-day facility in Texas City, Texas, exploded, killing fifteen workers and injuring more than 180. Investigators found that the company had laid off dozens of inspectors and maintenance workers in the run-up to the disaster, saving $1.2 million in annual expenses. The company was cited for more than three hundred safety violations at the refinery and fined $21.3 million.

That wasn't the end of it. In October, 2009, the Occupational Safety and Health Administration fined BP a record $87.4 million for 709 safety violations at Texas City, including a failure to repair many of the kinds of problems that led to the fatal 2005 blast. In fact, between 2007 and February, 2010, two BP refineries—Texas City and another in Toledo, Ohio—received 862 OSHA violations, ninety-seven percent of the industry total for flagrant violations, according to an analysis by the Center for Public Integrity, a non-profit Washington watchdog group. Of those, 760 were classified as "Willful egregious," according to the center, which obtained the records under a request supported by the Freedom of Information Act.

"The only thing you can conclude is that BP has a serious, systematic safety problem in their company," Jordan Barab, Deputy Assistant Secretary of Labor for occupational safety and health, told the Center for Public Integrity. Similar conclusions were drawn on Capitol Hill.

"You had 760 violations in five years. Sunoco had eight safety violations. ConocoPhillips had eight . . . Citgo had two . . . and ExxonMobil had one safety violation the same time period you had 760," Rep. John Sullivan (R-Ok.), scolded Hayward at a June 17 hearing. "How the heck do you explain that?"

Hayward, who became BP's chief executive in 2007, said most of the problems predated his leadership, including a 2006 BP pipeline leak that resulted in the largest oil spill ever on Alaska's North Slope, spewing nearly 300,000 gallons of crude oil over frozen tundra near Prudhoe Bay.

"I think we acknowledged in 2005 and 2006 that we had

serious issues," he told members of the House Oversight and Investigations Subcommittee. "We have made major changes in the company over the last three to four years," he said. "I set the tone from the top by saying very clearly, safe reliable operations were our number one priority."

The 2005 Peach Bowl was a great football game for Louisiana State University fans. The Tigers routed the University of Miami Hurricanes 40-3 under the Georgia Dome. It was especially satisfying for one lucky Tiger from Lake Charles, Louisiana, who got to fly to Atlanta with his wife on a private plane to watch the game. There was, though, one catch. He was a federal inspector charged with ensuring the safe operations of offshore oil companies in the Gulf of Mexico. One of those companies paid for the trip.

When asked about the incident, "he explained that he was a 'big LSU fan,' and he could not refuse the tickets," federal investigators wrote in a March report issued by the inspector general's office of the Department of the Interior.

For most of the past decade, the Lake Charles office of the federal agency responsible for ensuring safe and responsible offshore operations has been compromised by one such instance after another. In a region where oil rigs are as familiar to outdoorsmen as cypresses and mangroves, where friendships that begin at childhood can endure for a lifetime, and where oil companies have taken pride, and profit, in working out problems largely on their own, the mission of government oversight became badly blurred.

At Lake Charles, inspectors who are supposed to scrutinize the operations of oil companies like BP have been treated to golf tournaments, hunting trips, fishing expeditions and skeet-shoots, compliments of the very companies they were paid to oversee. One inspector told investigators he had participated in at least a dozen such events, paid for by offshore oil and gas companies. He went home from one with a brand new shotgun, a "door prize," he told investigators.

Federal inspectors routinely accepted lesser gifts ranging from sports tickets and jackets to baseball caps and pocket knives. In at least one case, a federal inspector was hired by an oil company after weeks spent negotiating his future salary and benefits while he was inspecting that company's offshore operations. The overall result was an atmosphere of such laxity that oil company employees sometimes penciled in their own responses to official inspection forms, leaving it for federal inspectors to merely ink in those responses later, according to the report of the investigation.

The inspector general's office investigated the Lake Charles office of the department's Minerals Management Service, the now-defunct agency that for nearly three decades was responsible for oversight of offshore oil and gas operations. The inquiry followed an anonymous letter to the U.S. Attorney's office in New Orleans complaining that the close and sometimes casual ties between MMS and the companies it was supposed to be regulating had finally gone too far.

"For years, there's been a scandalously close relationship between oil companies and the agency that regulates them," President Obama said in a May 27 news conference at the White House. "I've heard people speaking about the dangers of too much government regulation," Obama said. "And I think we can all acknowledge there have been times in history when the government has overreached. But, in this instance, the oil industry's cozy and sometimes corrupt relationship with government regulators have meant little or no regulation at all." Concluded Obama, "The oil and gas industry has leveraged such power, that they have effectively been allowed to regulate themselves."

Much like the revolving door that links elected officials and government bureaucrats to the oil lobby in Washington, MMS workers moved fluidly from the oil industry to the regulatory agency, and sometimes back again, in a professional community that bordered on fraternal.

"Obviously we're all oil industry," the district's manager explained to investigators. "We're all from the same part of the country. Almost all of our inspectors have worked for oil companies out on these platforms. They grew up in the same towns. Some of these people, they've been friends with all their life. They've been with these people since they were kids. They've hunted together. They fish together. They skeet shoot together."

Lifetime bonds of friendship and community are central to the culture in south Louisiana and, indeed, along much of the Gulf coast. MMS had allowed that to permeate much of its operations in the region, resulting in what acting inspector general Mary Kendall called "egregious misconduct" by inspectors.

"Of greatest concern to me is the environment in which these inspectors operate—particularly the case with which they move between industry and government," Kendall wrote in a letter to Secretary of the Interior Ken Salazar. "The individuals involved in the fraternizing and gift exchange—both government and industry—have often known one another since childhood."

With agency fecklessness highlighted by the Macondo blowout, Obama directed Salazar to break up the agency into three distinctive parts, reflecting three separate missions, as my organization, the Natural Resources Defense Council, had urged. The agency was renamed the Bureau of Ocean Energy Management, Regulation and Enforcement.

As the new name suggests, the old MMS was saddled with separate, and sometimes conflicting, assignments. To manage offshore oil production, the agency was required to lease sites, called blocks, where drilling and production can take place, and to assess royalty payments based on the value of oil and gas produced offshore. That generates revenue for the federal government—projected to be $100 billion over the coming decade, according to a July analysis by the non-partisan Congressional Budget Office. Those revenues combined with the national security objective of increasing U.S. production of domestic oil to create a powerful institutional incentive for timely approval of up to one thousand drilling permits a year in the Gulf.

The regulation part of the MMS role had more to do with developing specific standards and guidelines for offshore oil operations. In an industry with a long history of finding its own way into increasingly difficult waters, an industry rooted in a largely unregulated past, government safeguards have often lagged industry practices and technology by years. And, given the level of specialized expertise required to develop emerging offshore technologies, the regulators relied on the industry in ways that were so extensive and pervasive that it amounted, at times, to self-policing. The agency has looked to the industry for at least seventy-eight standards it has adopted as regulations, the Washington *Post* reported August 25, 2010. The article documented the "partnership" that existed between the agency and the industry it was supposed to oversee. "That path didn't work, and the public got let down in an enormous way," said Tom Strickland, chief of staff to Secretary of the Interior Ken Salazar. "There is now agreement—whether everyone in the industry agrees or not, because it's coming, it's happening—that we need more oversight, more regulation," Strickland told the *Post*.

In the past, though, both industry officials and regulators have expressed pride over the outcome of the arrangement, pointing out that Gulf offshore wells go from discovery to production in sixty-eight months, on average, compared to eighty months worldwide and 116 months in Europe. The regulation, though, hasn't kept pace with the industry, said William Reilly, co-chair of the National Commission on the BP Deepwater Horizon Oil Spill and Offshore Drilling.

"The industry has transformed itself into a technological behemoth," Reilly told *The Washington Post* in August, "And the government regulator is a ninety-eight-pound weakling."

Enforcement was yet a third and separate role, in conflict, at times, with the need to manage federal revenue from the wells. A well that's been shut-down, after all, can cost the government money in lost royalties. The agency lacked the resources, moreover, to effectively enforce safeguards in the Gulf, where there were sixty inspectors responsible for overseeing 4,000 offshore platforms. "This is juxtaposed with the Pacific Coast, which has ten inspectors for twenty-three facilities," Kendall told the House Subcommittee on Energy and Mineral Resources in a June 17 hearing. Inspectors, at least in the Gulf, operate independently, "With little direction as to what must be inspected or how," said Kendall. Training for inspectors is willy-nilly—"primarily on-the-job," as she put it. Finally, she suggested, penalties don't fully reflect the risk.

"We question," said Kendall, "Whether the civil penalty regulations are tied appropriately to the seriousness of the violation and the threat to human safety, property and the environment. Again, the regulations are sparse."

Given the array of structural, cultural and logistical shortcomings the agency faced, it's surprising, in some ways, that it took this long for a Macondo-like catastrophe to occur. The industry takes credit for its successes, but Chris Oynes, a former senior MMS supervisor in the Gulf, said federal inspectors are due tribute for striking the right balance between promoting the oil production the country needs and protecting public health and safety while safeguarding a unique national resource.

"That is a very noble mission," Oynes, who resigned a month after the Deepwater Horizon disaster, told *The New York Times.* "I thought we had done a pretty good job of addressing the challenges that come with deep water," he said in an article published August 8. "My opinion has not changed." Similar views have been expressed by other career MMS officials, who point out that the vast majority of inspectors work hard, at great sacrifice, on what they consider vital work that must be done right.

"I can tell you without hesitation that everyone in that regulatory program is fully committed to safety and pollution prevention—inspectors, engineers, geologists, scientists and others," said Elmer Danenberger, former chief of the offshore regulatory program at MMS. "And the inspectors, they expose themselves to considerable risk every day when they fly offshore and they go around platforms—every day," Danenberger told the Senate Energy and Natural Resources Committee at a May 11 hearing. "After Hurricanes Ivan, Katrina, Rita, Gustav, Ike, even when their own personal lives were disrupted, these people were on the job the next day doing everything they could to get production restored in a safe and timely matter. And ethics—these people won't take a donut from industry. I know. I've tried to set them up."

Because BP had distinguished itself as one of the worst offenders in the industry, though, regulators should have been even more vigilant, said Jeanne Pascal, who spent eighteen years as a Seattle-based attorney for the Environmental Protection Agency. "There comes a point in time where we say enough is enough," Pascal told *The Seattle Times* in June.

Beyond the shortcomings in federal oversight of offshore drilling and production operations, there is the separate matter of public overview of oil company plans for responding to a catastrophic oil spill. As it turned out, BP had no effective way to put a stop to the gusher. It took three months for the company to cap the well and another month to kill it permanently.

"It was entirely fair criticism to say BP dropped the ball when it came to planning for a major leak," Hayward told the *Financial Times* in an article published June 3. The company, he said, "did not have the tools you would want in your tool-kit." Nor did BP have, as it had assured regulators in its Macondo permit application, an effective plan for containing the oil from a catastrophic spill and protecting one of the richest, most ecologically diverse and biologically productive places anywhere on Earth from the ravages of toxic crude.

Why not?

Under the provisions of a law Obama said "Was tailored by the industry to serve their needs instead of the public's," regulators have just thirty days to decide whether to approve an application to drill in the Gulf. "That leaves no time for the appropriate environmental review," Obama said. "The result is, they're continually waived." MMS routinely used what it called a categorical exclusion to exempt applicants from the rigors of an extensive environmental review. From the standpoint of environmental protections, the applications were rubber-stamped.

"You just automatically gave the environmental waiver," said Obama, "because you couldn't complete an environmental study in thirty days." The result of that approach is evident in BP's application for the Macondo well—Mississippi Canyon Block 252 is its official designation. In category after category, real risks to habitat and marine life are dismissed with a cursory mention and boilerplate assurances of minimal threat easily managed.

"An accidental oil spill that might occur as a result of the proposed operation in Mississippi Canyon Block 252 has the potential to cause some detrimental effects to fisheries," the application reads. But then it concludes, "no adverse impacts to fisheries are anticipated as a result of the proposed activities." In fact, the spill caused a quarter of American Gulf waters to be closed to fishing, throwing thousands of watermen out of work.

What about sea turtles?

"Oil spills and oil spill response activities are potential threats that could have lethal effects on turtles," the application states. "No adverse impacts to endangered or threatened sea turtles are anticipated as a result of the proposed activities." Sea turtles—including Kemp's ridley, which are endangered—have washed up dead from the oil.

Can it kill a pelican?

"Birds could become oiled. However, it is unlikely that an accidental oil spill would occur from the proposed activities. No adverse impacts to marine and pelagic birds are anticipated as a result of the proposed activities." Hundreds have been found dead or dying.

Damage to beaches?

"An accidental oil spill from the proposed activities could

cause impacts to beaches. However, due to the distance to shore-—forty-eight miles—and the response capabilities that would be implemented, no significant adverse impacts are expected." The oil soaked some 630 miles of beaches.

BP offered the same hollow assurances for the wetlands, shore birds, coastal nesting grounds, wildlife refuges and wilderness areas. And, to make sure it covered all its bases, BP provided a blanket assurance. "In the event of an unanticipated blowout resulting in an oil spill, it is unlikely to have an impact based on the industry-wide standards for using proven equipment and technology for such responses, implementation of BP's Regional Oil Spill Response Plan, which address available equipment and personnel, techniques for containment and recovery and removal of the oil spill."

Don't worry, BP assured the regulators, everything will be fine. Trust us. And, for the most part, they did. "It is this kind of blind faith . . . that has led us to disaster," Representative Edward Markey (D-Mass.), chairman of the Energy and Environment Subcommittee, said at a June 15 hearing. "We now know the oil industry and the government agency tasked with regulating them determined that there was a zero chance that this kind of undersea disaster could ever happen."

That was reflected, Markey said, in the subcommittee's analysis, based on information provided by the major oil companies, that the industry piled up $289 billion in profits over the previous three years, spent $39 billion to explore for and produce new oil and gas wells, yet invested, on average, "A paltry $20 million per year" on research and development for safety, accident prevention and oil spill response. That's less than one-tenth of one percent of the industry's profits that went toward trying to ensure an adequate response to an accident.

As a result, while the technology and expertise for deepwater drilling had advanced several generations in just twenty years, there had been little if any substantive improvements in oil spill response since the *Exxon Valdez* disaster in 1989. The same toxic dispersant—Corexit—was used, despite the unanswered questions about the harm it would do to habitat and wildlife.

"What I see is a primitive response industry . . . to me, it differs very little from what I saw twenty years ago," said Reilly, who worked closely on the *Exxon Valdez* aftermath as Administrator of the Environmental Protection Agency in the administration of George H. W. Bush. "No one has put money into research and to upgrade it . . . You would have had to have government leadership and pressure to address it. We just didn't have it."

Who, then, will protect the public? Industry does not do that on its own. That is the job of government. That is exactly why we organize ourselves, as a nation, in a way that brings collective resources to common goals. Who else is looking out for us, the public?

Scapegoating BP is not the answer. Because, as it happens, BP's oil spill response plan was identical—in many cases verbatim—to the plans filed by Exxon, Shell, Chevron and ConocoPhillips for applications to drill for oil in the Gulf. Each of these companies filed with regulators an off-the-shelf response plan prepared by a company that specializes in such plans.

"These are cookie cutter plans," Representative Henry Waxman (D-Calif.), said at the June 15 hearing. "ExxonMobil, Chevron, ConocoPhillips and Shell are as unprepared as BP was, and that's a serious problem."

The plans were so general that they included references to protecting walruses, an animal that hasn't been within three thousand miles of the Gulf since the last Ice Age—if ever. And yet, the plans included little, and in some cases nothing, about hurricanes.

"It's unfortunate that walruses were included," Exxon chairman Rex Tillerson told Markey's subcommittee, with executives from the other companies mentioned above at his side. "It's an embarrassment."

Since each company filed, in essence, the same oil spill response plan, there's no reason to believe the other companies were any better prepared than BP to confront a catastrophic spill in the Gulf.

"When these things happen, we are not well equipped to deal with them," Tillerson conceded. "That's why the emphasis is always on preventing these things from occurring, because, when they happen, we're not very well equipped to deal with them. And that's just a fact of the enormity of what we're dealing with."

In other words, "no matter which one of the oil companies here before us had the blowout, the resources are not enough to prevent what we're seeing day after day on the Gulf," asked Congressman Bart Stupak (D-Mi.). "That is correct," Tillerson replied.

"So, but for the grace of God there goes I, right? It's BP this time; it could be ExxonMobil tomorrow; it could be Chevron tomorrow . . . 'We satisfied the application, but, in reality, we can't respond to a worst-case scenario,'" said Stupak. "A response is underway. It is having some effect," Tillerson said, referring to the efforts by BP and the Coast Guard to contain the oil. "But there is no response capability that will guarantee you will never have an impact. It does not exist. And it probably will never exist."

The trusting, sometimes breezy and almost laissez-faire attitude toward public oversight of the offshore oil industry is no mere political accident or legislative aside. It is the product of a deliberate, concerted and well-funded decades-long assault on a fundamental principle reflecting core American values. That principle is this: when an activity or industry puts at risk public health and safety, or the environmental resources we all depend upon, protections are not only legitimate, they're essential, to safeguarding the collective good. They are the only option, and they are the only way to spread the costs of public oversight into the price of production in a fair, rational and equitable way.

For the first two hundred years of this country's existence, those protections were minimal and hard to pin down. Informed by the sensibilities of naturalists dating to Henry David Thoreau and advanced by presidents like Theodore Roosevelt, notions of public defense of our common inheritance breathed, sometimes barely, within a frail and nebulous body of common law precedent and local codes. Together, though, they were inconsistent, unevenly applicable from one area to another and subject to capricious interpretations and findings in court. Water and air pollution, for example, were considered to be local issues, first regulated by municipal governments and later by state governments. Toxic chemicals were rare before World War II, and could be addressed site-by-site.

As it became ever clearer that pollution crossed city and state boundaries, however, and as our economy became more closely linked among states, Americans began demanding a more predictable, nationwide and comprehensive approach. Throughout the 1960s, largely galvanized by Rachel Carson's book *Silent Spring*, which documented the dangers of synthetic pesticides, pressure for a larger and more active federal role grew. At first the federal government role focused on research. As companies became larger, though, and states proved unable or unwilling to impose adequate protections, demand for direct federal engagement increased. It all came together in 1969. That year, an offshore oil blowout spewed four million gallons of crude oil into the waters off the coast of Santa Barbara, California, and pollutants ran so freely through the nation's industrial heartland that Ohio's Cuyahoga River caught fire. Those and other environmental wake-up calls helped to galvanize a broad-based bipartisan consensus that forged a set of bedrock protections that have defended public health, safety and the quality of our environment for a generation and more.

"The destiny of our land, the air we breathe, the water we drink is not in the mystical hands of an uncontrollable agent, it is in our hands. A future which brings the balancing of our resources—preserving quality with quantity—is a future limited only by the boundaries of our will to get the job done."

The words of some new-age, pie-in-the-sky idealist? Hardly. The speaker was President Richard M. Nixon in a radio address to the nation from the Oval Office on Valentine's Day, 1973.

For Nixon, under siege at the time by congressional Watergate investigations that would lead to his resignation the following year, the words were more than mere lip service. Nixon signed some of the most far-reaching environmental legislation in our history, beginning with the National Environmental Policy Act of 1969, the landmark bill that began to organize a comprehensive government response to the mounting threats pollution posed.

"Our national government today is not structured to make a coordinated attack on the pollutants which debase the air we breathe, the water we drink, and the land that grows our food," Nixon said in a July 9, 1970 message to Congress. "The government's environmentally-related activities have grown up piece-meal over the years," he said. "The time has come to organize them rationally and systematically."

Introduced in the Senate by Henry Jackson, a Democrat from Washington, and in the House by John Dingell, a Michigan Democrat, the National Environmental Policy Act made protections from risks to human health and the environment the basis for a new body of federal law. It made environmental impact statements a requirement of major federal actions. It created the Environmental Protection Agency, with broad authority to enforce prohibitions and orders deemed essential to advancing environmental policies and goals. And it established the White House office of the Council on Environmental Quality, to provide policy analysis and advice to the president on environmental issues.

So strong was bipartisan support for the bill that it passed unanimously in the Senate and by the overwhelming vote of 372-15 in the House. "The 1970s must be the years when America pays its debt to the past by reclaiming the purity of its air, its waters, and our living environment," Nixon said of the legislation. "It is literally now or never."

The first Earth Day demonstrations, in April of 1970, helped build momentum for the passage of the Clean Air Act of 1970. Before that, bits and pieces of early clean air legislation had been passed—including in 1963, 1965 and 1967—that were weak and uncoordinated. The 1970 act helped to define the modern air pollution control program. It set up a system of science-based air pollution standards and set limits on emissions from vehicles and factories.

Written by Senator Ed Muskie the bill passed the Senate

with unanimous backing from both sides of the aisle. The House version dropped auto emission reductions deadlines, but they were restored in compromise legislation that Nixon signed.

In February, 1971, Nixon sent a letter to Congress, outlining an ambitious environmental agenda that was sweeping in scope. At a time when cars were releasing more than 200,000 tons of lead into the atmosphere each year, he proposed reducing the lead in gasoline. Two years later, he took action to do just that. He called for strengthening pesticide controls, and later signed legislation that accomplished that. He called for, and his EPA later set, the first standards for reducing sulfur dioxide emissions, a program that eventually reduced those pollutants dramatically. Nixon's letter, in fact, amounted to an environmental manifesto, setting policy goals on subjects running the gamut from toxic chemical use to ocean dumping.

"The battle for a better environment can be won," Nixon said, "And we are winning it." The next year, Nixon vetoed the Clean Water Act of 1972, citing cost concerns. Congressional support was strong enough—and bipartisan enough—to secure the two-thirds majority vote required in each house to override the president's hand. He signed it into law three weeks before Election Day. The new law put an end to the dumping of raw sewage, authorized billions of dollars to clean up municipal waste around the country and required all polluters to use "best available technology," as established by the EPA, to reduce pollution. In sweeping language, it aimed to restore the chemical, biological and physical integrity of our waters to make them all fishable and swimmable.

In 1973, Nixon directed Russell Train, chairman of the Council on Environmental Quality, to write legislation that would become the Endangered Species Act. It established specific protections and conservation requirements for wildlife threatened with extinction, including penalties for killing such animals. It passed in the Senate 92-0 and in the House 355-4.

"Nothing is more priceless and more worthy of preservation than the rich array of animal life with which our country has been blessed," Nixon said when he signed the legislation in December 1973. "It is a many-faceted treasure of value to scholars, scientists, and nature lovers alike, and it forms a vital part of the heritage we all share as Americans . . . a heritage we hold in trust to countless future generations of our fellow citizens."

These and other touchstone pieces of environmental legislation, passed with overwhelming bipartisan support, signed by President Richard Nixon and enacted into law, laid the groundwork for essential safeguards that have created a cleaner, safer and healthier environment for us all.

In more recent decades, however, the entire edifice began to come under assault, part of an assertive broadside against government oversight and regulation of industries and activities that put the public at risk. Indeed, the movement took aim at government itself, and the career professionals who make it work.

"Government is not the solution to our problem; government is the problem," President Ronald Reagan declared in his inaugural address on January 20, 1981. "It is my intention," said Reagan, "To curb the size and the influence of the federal establishment."

The sentiment meshed smoothly with an election campaign he'd waged the year before, during which Reagan attacked environmental regulations and "extremists" at the EPA, blaming both for the slow growth and high unemployment he inherited from President Jimmy Carter. While running against Carter in 1980, Reagan famously asserted that "Trees cause more pollution than automobiles do," prompting an opponent to show up at one of his rallies to post this message on a nearby tree: "Chop me down before I kill again." As president, Reagan expressed exasperation with what he called "environmental extremism," once telling reporters "I don't think they'll be happy until the White House looks like a birds' nest."

Beginning with the appointment of James Watt as Secretary of the Interior, Reagan set a tone that encouraged the rollback, and even the denigration, of essential environmental protections. Watt drew congressional fire when he proposed opening a billion acres of coastal waters to offshore drilling on the Outer Continental Shelf along most of the Atlantic, Gulf and Pacific coasts. The plan was later scaled back. He aided the coal industry by shutting down offices of the regulatory agency that oversaw strip mining. He also had ninety-three percent of the rules of the Office of Surface Mining rewritten and watered down in ways that favored industry, *Time* magazine reported in October, 1983. The article assessed Watt's impact when the pugnacious cabinet member stepped down after ridiculing affirmative action.

There were other instances in which the Reagan administration took aim at public protections and environmental safeguards. Early on, for example, Reagan slashed the staff of the White House Council on Environmental Quality to just eight, down from fifty, according to a National Public Radio assessment following Reagan's death in 2004. And Reagan's EPA exempted wastes generated by the oil and gas industry from coverage under the Resource Conservation and Recovery Act of 1976. That law gave the EPA authority to control hazardous waste during its entire life cyclen'tfrom the production of such waste, in other words, to its disposal. Waste from oil and gas production often contains heavy metals and toxic chemicals. Its exemption was a serious blow to the protection of public health, safety and the environment.

Reagan's first EPA administrator, Anne Gorsuch, reduced the agency's budget by twenty-two percent. Clean water regulations were spelled out in a book six inches thick when she took office: she quickly cut it to a half-inch booklet. She resigned after less than two years in office when Congress cited her for contempt, after she refused, on Reagan's orders, to turn over documents related to a scandal over the mismanagement of $1.6 billion in funds meant to help clean up toxic spill sites.

Perhaps the worst of the Reagan legacy, though, was simply work left undone. Badly needed enhancements and extensions to protections for air, land and water went largely unaddressed or underfunded. Congressional deadlines in environmental laws went unmet, enforcement plummeted and new science was ignored. It was, in the minds of many, eight years lost when we could ill afford lost time.

In 1989, the *Exxon Valdez* oil tanker ran aground and dumped eleven million gallons of crude oil into Prince William Sound in Alaska after the ship's captain, an alcoholic, hit the bottle. Until the Deepwater Horizon disaster, it was the worst oil spill in the nation's history. In response, Congress passed the Oil Pollution Act of 1990 without a single vote of dissent. The bill passed in the Senate 99-0, with one member not voting, and in the House 360-0, with seventy-two members choosing not to vote rather than being on record against it.

President George H. W. Bush, who'd once run his own oil company in Texas, signed the legislation with reservations. "The Act addresses the wide-ranging problems associated with preventing, responding to, and paying for oil spills," he said at the time, going on to praise the environmental safeguards in the bill, such as improved standards for oil tanker construction, crew licensing and manning.

Bush signed the bill just sixteen days after Iraqi troops invaded neighboring Kuwait over an intractable dispute involving oil. He had already begun deploying U.S. forces to the Persian Gulf as part of a massive buildup in advance of Operation Desert Storm, the U.S.-led war to evict Iraq from Kuwait.

He took issue with a provision in the bill calling for a moratorium on offshore oil and gas exploration off the coast of North Carolina. "It is shortsighted," said Bush, because that oil and gas "could be used to offset our dependence on foreign energy sources." Said Bush, "Such a moratorium is ill-advised in view of recent events in the Persian Gulf, where I have found it necessary to deploy American soldiers 7,000 miles from home to protect our vital national interests."

Yet Bush did push back against some of the Reagan-era offshore policies. He cancelled a host of scheduled lease auctions, bought back federal leases off the Florida coast, and established a marine sanctuary in California's Monterey Bay. As part of his effort to "reinvent" government, President Bill Clinton pursued a policy of "performance-based regulation", effectively increasing the oil industry's influence over government oversight. MMS staff was cut as part of larger shift toward partnership between the agency and the industry it was authorized to oversee.

The real work of rolling back public protections for the good of the oil industry fell to Bush's son, President George W. Bush, and his vice president, Dick Cheney. Both made it a priority to increase domestic oil production, for economic and national security purposes. Cheney, in fact, made it a personal mission.

As Secretary of Defense under the elder President Bush, Cheney oversaw U.S. efforts to drive Iraq from Kuwait. In 1995, he began working for the Halliburton Company, where he was president and chief executive officer. The Dallas-based international oil services company was attracted to Cheney's global Rolodex and his long experience in government. Before his Pentagon assignment, Cheney was chief-of-staff to President Gerald Ford and served five terms in the House of Representatives.

As vice president under George W. Bush, Cheney led a secretive energy task force that relied almost exclusively for outside input on corporations that produce oil, gas, coal and electricity. A seasoned Washington insider expert at federal bureaucratic machinations, Cheney carefully tailored the National Energy Policy Development Group to avoid sunshine laws requiring public disclosure of task force proceedings. He refused requests to make public the minutes from task force meetings, or even to reveal the names of industry sources that met with him or his staff.

It wasn't only environmentalists who were rebuffed. Cheney asserted executive privilege to stonewall even the non-partisan Government Accountability Office, claiming there were Constitutional issues of separation of powers at stake. He took the dispute to the U.S. Supreme Court, which ruled in his favor in 2004, finding that the GAO did indeed lack the authority to compel his task force to make public its work.

The substance of it, though, became clear enough when the task force issued its National Energy Policy Report, the blueprint for the Energy Policy Act that was approved by the Republican-controlled Congress in July, 2005.

"This legislation promotes dependable, affordable, and

environmentally sound production and distribution of energy for America's future," Bush said in a prepared statement when he signed the bill into law the next month. The law provided billions of dollars in tax breaks for oil, gas, nuclear power and coal companies, and gave short shrift to alternative and renewable forms of energy. Indeed, one of the first things Bush did, in his first week in office, was to try to rescind rules increasing the efficiency of air conditioners—which use a third to a half of peak electricity in much of the country. Several states and the Natural Resources Defense Council successfully opposed this in court.

Despite Cheney's best efforts to conceal proceedings that would profoundly impact the rules of the road for an industry whose tentacles reach deep into every quarter of American economic life, the NRDC was able to secure documents that provide a telling glimpse behind the curtain of secrecy that shrouded the group's work.

In the spring of 2002, a federal judge ordered the Department of Energy to release to the NRDC 13,500 pages of documents related to task force proceedings. While they were heavily censored, the documents revealed that Cheney's task force had sought extensive advice from utility companies and executives in the oil, gas and coal industries, and then incorporated their recommendations, often word for word, into the energy plan.

The documents show, for example, that Bush administration Energy Secretary Spencer Abraham held eight meetings with energy and business leaders on the plan as the task force was drafting its report. Meanwhile, he held no meetings with environmental leaders or advocates for energy efficiency or renewable sources of power and fuel. High on the list of inside-the-moat advisers were executives and lobbyists from BP, Exxon,

ConocoPhillips, Shell, Chevron and the industry trade group, the American Petroleum Institute (API).

The task force's recommendations covered far more than just oil drilling. One series of recommendations, which became the basis of numerous lawsuits and legislative battles, proposed weakening air pollution standards for coal-fired power plants. Emissions from such sources cause acid rain and smog, contributing to the premature deaths of more than twenty thousand people each year. Not surprisingly, these proposals were similar to industry recommendations.

Many recommendations did, however, address oil specifically. One telling example in the documents obtained by the NRDC came in an email sent on April 10, 2001, by oil industry lobbyist Kyle Simpson to Joseph T. Kelliher, a member of Cheney's task force, a senior adviser to Abraham and later the chairman of the Federal Energy Regulatory Commission.

In the "Dear Joe" email, Simpson provided draft legislation focusing on ultra-deepwater production research and development, and called for subsidies for the industry to help it drill in the Gulf of Mexico and tap rich deposits of oil and natural gas. "Here are three brief documents describing the Ultra-deepwater projects that we discussed last week. This would be a very important project for our energy future," he wrote. Simpson called on the energy secretary to "develop and implement an accelerated cooperative program of research and development, develop natural gas and oil reserves in the ultra-deepwater off the Central and Western Gulf of Mexico."

"The largest oil and natural gas resource ever discovered in the United States lie beneath the ultra-deepwater of the Central and Western Gulf of Mexico," wrote Simpson. "The pace of development of the resource and the realization of ultimate contribution that can be made to the energy security of the United States is impaired by the enormous cost and physical challenges to the development of these resources. Lowering the cost and improving the efficiency of developing this new resource must be a priority of our national energy policy."

The phrase "lowering the cost" is the kind of code industry often uses in arguing for relaxed environmental and safety protections or speeding the review of permits, exactly as happened with offshore drilling applications.

Simpson is now the policy director at a Washington law firm. Its website describes him as "One of the principal architects of the Ultra-deepwater and Unconventional Onshore Natural Gas and Oil Research and Development provision which was included in the Energy Policy Act of 2005."

Kelliher also was in close contact with James Ford, an official with the API, the oil industry trade group. Ford sent Kelliher a March 20, 2001, email with copies of the API's position papers and a "suggested executive order to ensure that energy implications are considered and acted on in rulemakings and executive actions."

On May 18, 2001, Bush issued an executive order with similar wording to the draft supplied by API. The president's order required all agencies to study energy effects when preparing regulations and then submit the findings to the federal Office of Management and Budget. The directive required that agencies avoid "Any adverse effects on energy supply, distribution, or use," and develop "reasonable alternatives to the action with adverse energy effects."

In other words, the executive order sent a clear and unmistakable signal to all relevant agencies, such as EPA, the National Oceanic and Atmospheric Administration or MMS, that they should not a adopt any regulations that impede the energy industry or oil and natural gas production—an approach that melded nicely with the energy task force recommendations and the 2005 energy law adopted by Congress. Federal regulators overseeing offshore oil development in the Gulf of Mexico appear to have read the memo.

In the same March document sent to Kelliher, Ford recommended that the administration reaffirm the primary authority of the Minerals Management Service to regulate all offshore oil and gas leasing, exploration, development and production activities. In addition, the API official strongly recommended enactment of "deepwater royalty relief" to save the industry billions of dollars.

"To encourage investment in domestic oil and gas resources on the Outer Continental Shelf, Congress enacted the Deepwater Royalty Relief Act of 1995 to suspend the payment of royalties for specific initial quantities of oil and gas produced from the OCS in water depths greater than 600 feet," the API document said. "This incentive was very successful and resulted in billions of dollars in additional revenue to the United States and a significant increase in oil and natural gas production from OCS waters."

On February 1, 2001, Energy Secretary Abraham, Cheney and the energy task force also heard from the American Association of Petroleum Geologists and its president, G. Warfield "Skip" Hobbs. In a policy statement, Hobbs assured the task force that offshore oil and gas resources could be developed in the Gulf of Mexico in an "environmentally responsible manner, with no lasting harm," owing in part to the "technological advances" the industry had made and its ability to mitigate environmental damage.

The NRDC also uncovered evidence showing the Bush administration implemented pro-industry energy policies requested by Chevron. The company provided several recommendations, ranging from easing federal permitting rules for energy projects to relaxing standards on fuel supply requirements, which ultimately were included in the president's national energy plan.

In a February 5, 2002, letter to President George Bush and copied to Energy Secretary Abraham, Chevron CEO David J. O'Reilly recommended four short-term actions the administration should take to "eliminate federal barriers to increased energy supplies."

"Substantial federal policy and regulatory barriers constrain the supply of U.S. natural gas and crude oil. They restrict or prevent responsible energy development on most of the Outer Continental Shelf and in many highly prospective areas of Alaska and the Rockies," according to O'Reilly.

"Government must improve resource access, streamline application and permit processes, eliminate unnecessary delays and reject unjustified opposition to new energy leasing and development," O'Reilly recommended.

The energy task force included all of Chevron's recommendations in its report, the document that defined an eight-year effort to tilt the field of public protections in favor of the big oil companies and other energy concerns. That was the pay-off, as far as industry was concerned, for a decades-long assault on the political consensus supporting needed environmental safeguards.

"That movement sought to turn forty years of bipartisan environmental protection on its head, and it did," Bob Marshall, outdoors editor for *The Times Picayune* of New Orleans, wrote in a column published May 23. "If Louisiana is lucky, the Deepwater Horizon will be the Three-Mile Island of deep-ocean drilling. It will be the event that demonstrates to all who sneer at environmental regulation just what's at stake here, and the enormous danger facing our coast each and every day."

In the fall of 2008, seven years after the well was discovered, Blind Faith began producing its first oil. Now owned by Chevron and a second partner that has replaced BP, the operation is delivering nearly three million gallons of oil each day, and is expected to do so for the next twenty years, an achievement highlighted in a corporate video. "As people demand more and more energy," a narrator intones, "We must look farther and wider to find it."

That's exactly what Jason Anderson and ten of his co-workers were doing the night they were killed aboard the Deepwater Horizon. They were men of grit and courage, of hope and faith.

"Democracy rests on faith," President Lyndon Johnson felt compelled to remind the nation in 1965, when the country was riven by civil rights tensions, economic uncertainty, perilous global rivalries and the Vietnam War. "We are a nation of believers," he said. "We are believers in justice and liberty and in our own union. We believe that everyone must some day be free. And we believe in ourselves."

We've believed also that we can safeguard each other from harm. Now and then tragedy strikes in a way that can force us to look anew at the need to do that together, not by hope and belief alone, not blinded by our faith, but by deliberately learning the lessons of our own history and putting in place the protections we need.

"We want to prevent this from ever happening again," Anderson's mother Shelley said, standing in the halls of the Capitol not long after Deepwater exploded. "We don't want anyone to ever have to lose a husband or have to tell a five-year-old that her daddy's gone to heaven. Ever again."

Chapter 5

A Healthy Gulf

From the sparse shade of the Bimini top flapping over his sleek bay skiff, Sal Gagliano looked out on brackish waters where the Mississippi Delta makes its uneasy peace with the Gulf of Mexico. Still as a stone, a great blue heron stalked the shallows at the edge of a marsh, cradle to fingerling porgies, menhaden and trout.

"We are a nursery ground," said Gagliano, who has fished these waters since he was a boy, "one of the most productive estuaries that could ever exist."

He throttled up and skimmed out along the broad lip of Barataria Bay, veered sharply into the sun, and motioned toward a long line of dead cypress trees, skeletal trunks silhouetted in the haze like some distant funeral train. "It's because we're losing the marsh," he shouted above the engine's whine. "All the saltwater's killed them."

He banked into a wide canal, and glanced back over his shoulder. "Everything out there we just rode through used to be land," he said. "It's gone, G-O-N-E, and it ain't coming back."

He cut the engine and let the boat drift toward the edge of

an oil-soaked marsh. In its still-vibrant center, lime-green grass stood straight and tall, its yellow tips forming a willowy crown, golden in the morning sun. Along the water, though, reaching into the marsh ten feet or more, was a swath of coal-black waste where the oil had come, the grass gone dead and fallen, as if scorched by some unseen fire.

"We sacrificed something that shouldn't have been sacrificed," Gagliano said in somber eulogy to the marsh. "They'll have to show me they can re-grow the grass before I'll believe it, before the mud all erodes away."

The great marshes of Louisiana, home to forty percent of the nation's tidal wetlands, are dying: sinking and washing away. Over the past five decades, Louisiana has lost, on average, thirty-four square miles of wetlands each year—the equivalent of sixty football fields each day—according to the U.S. Geological Survey. That is the highest rate of erosion anywhere in the country, and possibly the world. Since 1932, Louisiana has lost enough wetlands to cover the state of Delaware. It is on track to lose at least that much again over the next fifty years.

Sea level is rising, wetlands eroding and the sediments that might naturally replenish the marsh don't flow there anymore. Along the lower delta, the Mississippi and all that it carries have been channeled within man-made earthen banks called levees. The levees are built to keep the river from flooding, but at the cost of starving the marsh of fresh silt. Elsewhere along the state's barrier islands and coastal wetlands, a similar story is told, made worse by decades of logging that has destroyed cypress swamps in exchange for the hardwood mulch that adorns golf courses, office parks and upscale suburban homes in communities far away.

As the wetlands have been weakened and winnowed out, the entire delta has become less resilient, increasingly susceptible to the ravages of wave action, currents and storms. And, as the cycle continues, with diminishing marshes to cushion the forces of raging hurricanes, the system is losing its ability to take nature's blows.

"We're washing away at a rapid rate of speed," said Ryan Lambert, who has watched the shocking loss of wetlands for three decades from the helm of a fishing boat. "When you've been in a place for thirty years and sometimes you're driving around and you don't know where you're at, something's not right."

The BP oil disaster didn't cause the problem, but it's made it worse. Toxic oil is killing the grass that holds marsh mud in place, leaving the front lines of scores of miles of fragile wetlands exposed to the currents and tides. The oil industry has exacerbated wetlands loss in other ways. Crisscrossing the marsh with hundreds of miles of pipeline, and cutting canals for supply boats and rigs, the oil companies have opened thousands of gaps for seawater to rush in, bringing salt that kills freshwater habitat, further accelerating wetlands loss.

"We've sliced and diced the coastal wetlands, because that's what we've let the oil companies do," said Aaron Viles, campaign director for the Gulf Restoration Network, a New Orleans environmental advocacy group.

The industry has taken its toll in other ways too. Leaks and spills from pipelines, storage tanks, platforms and other equipment routinely pollutes regional waters and wetlands. Toxics Targeting Inc., a consulting firm in Ithaca, New York, assessed a sampling of federal spill response records and found scores of individual spills that together have dumped more than two hundred thousand gallons of oil and related industrial fluids into the Gulf and adjacent wetlands over just the past two decades. That doesn't count the mayhem that followed Hurricane

Katrina, when more than eight million gallons of oil spilled into the region as the result of damage to production, storage and processing facilities.

On the state level, Viles said, the industry is a political sacred cow. Political leaders have struck a bargain with the oil industry, he said, trading off the natural health of the region in exchange for jobs and tax revenue. "The politicians in Louisiana generally, with some exceptions, are wholly-owned subsidiaries of the oil companies down here and that's what we get: the Louisiana coastal ecosystem is in a state of crisis." Americans elsewhere, he said, have turned a blind eye, content to let the region bear the environmental brunt of the nation's oil habit.

"We're treated like a national sacrifice zone," Viles said in a telephone interview. "We've all driven the SUVs, we've flown in the planes and we've used the plastics, but we haven't all paid the environmental price," he said. "The Gulf of Mexico truly is ground zero for the environmental impact of our addiction to oil. We're getting hit first and hardest, and clearly the BP oil disaster is like an exclamation point in describing this."

Navy Secretary Ray Mabus hopes to change that line of thinking. Tapped by President Obama to prepare a comprehensive Gulf restoration plan, Mabus is a former Mississippi governor, a son of the Gulf who claims to understand the region, its unique culture and its people.

"I live here, I love this place," he said, explaining that his goal is to prepare a comprehensive Gulf restoration plan that addresses, not only the harm done this summer, but long-term ills as well. "Part of my job is not just the oil spill," said Mabus. "The oil spill has made things worse, but it didn't start it."

In July and August, Mabus held a series of meetings

around the region to hear directly from Gulf coast residents. In the summer of the accident, he came to Buras, a fishing hamlet along a narrow sliver of delta sandwiched between the levees that hold back the Mississippi River on one side and Barataria Bay on the other. When Hurricane Katrina made landfall, it slammed headlong into tiny Buras, bringing winds that topped one hundred fifty miles per hour and inundating the community with sixteen feet of water that stayed for forty-three days. In some ways, said Mabus, the oil spill is worse.

"This is not like Katrina," he told a crowd of three hundred people packed into the Buras auditorium on a steamy August night. "This is like a disaster that's just happening over and over and over again, and we've got to make sure that the rest of the country doesn't forget about the coast when the well's killed."

Mabus has been charged with developing two plans, really, one for dealing with short-term, or emergency, restoration needs, and the other for long-range efforts more comprehensive in scope. There's been no shortage of proposals and studies. Coastal restoration is an industry unto itself, imbedded in the region's academic, government and economic base.

Virtually every university and college in the region is involved in some kind of coastal management coalition, sometimes in league with non-profit conservation and environmental groups. There are more than three dozen companies devoted to coastal restoration and defense just in Louisiana, which was on track to spend some $3 billion a year on such projects before the BP oil disaster. With an eye on emerging opportunities worldwide, Greater New Orleans Inc., a regional economic alliance, has begun to bill the state as "the Netherlands of the U.S." based on its burgeoning expertise in lowcountry water management issues.

Louisiana passed coastal restoration legislation more than thirty years ago. In 1989, the state set up two agencies to oversee the protection and restoration of Louisiana's coasts and wetlands and to help coordinate state activities with those of the U.S. Army Corps of Engineers, the U.S. Geological Survey and other federal agencies that play a role in trying to manage coastal and wetlands issues.

The state has put together a ten-year coastal ecosystem restoration plan with proposals for fifteen specific projects, most of which would funnel river sediments back to wetlands. The projects could be put in place over the coming decade at an estimated cost of $2 billion. A proposal to restore shoreline along barrier islands near the mouth of Barataria Bay is being expanded to include repairs due oil-damaged coasts.

Mississippi passed its own coastal management legislation in 1972. It has a state coastal management and planning office, linking eighteen city and county partners. The group works with the National Oceanic and Atmospheric Administration on projects and plans ranging from wetlands dredging and filling to conservation of more than forty-four thousand acres of state and federal wildlife lands.

There are similar initiatives in Alabama, Texas and Florida. In 2004, the five coastal states formed the Gulf of Mexico Alliance to work cooperatively on ways to improve water quality, coastal habitat and the overall resilience of estuaries, wetlands and shores.

Mabus has pledged to use existing plans as a starting point, rather than try to reinvent the wheel. His challenge is to put together an overarching package of projects that make sense, can get done and can be paid for in a timely way. With government spending hampered by record debt, the Obama administration and Congress are looking at a number of funding options, including the use of offshore oil and gas royalties to help. Mabus has made clear, though, that what's happening to the Gulf, from the oil disaster and before, is not just a problem for those who live there. "This is a national issue," said Mabus. "Thirty percent of the oil, fifteen percent of the natural gas, a third of the seafood that we produce comes out of the Gulf. The rest of the country needs a healthy Gulf."

Much like the biological and energy wealth of the Gulf, its problems are rooted in history. As tectonic shifting opened up the Gulf basin two hundred million years ago, the pulling apart of continental plates thinned the Earth's crust at the sea's base, much like pulling warm taffy can leave it stretched too thin. As sediment piled onto the weakened ocean floor, the bottom sank, creating a deepwater bowl. The entire basin, though, is still undergoing slow-motion collapse, something geologists called subsidence. Sea level, meanwhile, is rising, as global climate change melts vast ice sheets and glaciers thousands of miles north and as warmer seas expand. Between the rising waters and the sinking base, the region is going under at the rate of about an inch every two and a half years.

"What you're seeing is that subsidence is taking place in conjunction with the absolute global sea rise," explained geologist Mark Kulp. "The land that was subaerial has now become subaqueous." Coastal Louisiana, in other words, is sinking, slowly swallowed up by the sea. "There's nothing we can do to stop the fact that the crust is sinking from the weight of the sediments on it," said Kulp, chairman of earth and environmental sciences at the University of New Orleans. "We just have to come up with protocol and plans to live with it."

Easier said than done. The Mississippi River is the region's lifeblood, irrigating farmland, nourishing aquatic habitat and marine life and providing a shipping conduit from Minnesota to the delta. Levees stretching nearly as far north as Memphis hold the river within artificial banks. An essential part of flood control, they also keep the river from finding its natural course and spreading its sediment along the broad deltaic plain. It is that sediment, though, that's needed to build and replenish the wetlands and barrier islands along the coast. The levees prevent that from happening.

"It's a starvation of the surrounding wetlands, essentially, that the levees are creating," said Kulp, who has studied the area for more than fifteen years. "It's just restricting the natural process that the delta needs to survive."

The levees aren't going anywhere, though, as long as industry and people remain on the low, flat delta. One idea being considered is to strengthen some of the wetlands by building spillways into the levees, to allow sediment to bleed into the marsh in places. That, said Kulp, might help to build more resilient marshes.

Along with sediments, the river brings something else: water enriched with nitrogen, the result of fertilizer runoff from farms and waste from sewage plants hundreds of miles upstream. The super-nitrogenated water soups up the formation of algae blooms, which then deplete the water of oxygen, creating so called "dead zones." In late August, more than five thousand dead crabs, sting rays, eels and fish were found near the mouth of the Mississippi, victims of low oxygen levels in the water, the Louisiana Department of Wildlife and Fisheries reported.

There's no quick fix for that problem, either. We need to strengthen, and then vigilantly enforce, restrictions on runoff in cities and communities far upriver. We need to figure out how to help farmers apply the fertilizer they need without generating excess runoff that goes into the Mississippi and harms communities downstream.

"If this were Martha Stewart's Vineyard, Cape Cod, San Francisco, people would be screaming," Gagliano said. "Help us out. Don't leave us like the black sheep." Behind that sentiment—widely held across the region—is fear of being abandoned as oil spill news drops off the front page.

"Out of sight, out of mind," said Mike Brewer, who was running for local office in Louisiana's Plaquemines Parish after a long career cleaning up hazardous waste. "If we're not in the media, we don't exist anymore."

Gulf coast residents have seen this movie before. Many felt forgotten after Katrina ripped through five years ago, leaving upturned lives and an estimated $100 billion of damage in its wake. "Everybody will forget about us," echoed Steve Gremillion, a marina owner on hard-hit Grand Isle, "and what is going to be left over here is a wrecked environment, an economy that is completely upside down and a very uncertain future for a lot of people."

Uncertainty isn't what first comes to mind at BP world headquarters in London. Subdued and understated, the building is identified only by a brown plaque the size of a coffee-table book with the letters "bp" in lower case and the courtly address: No. 1 St. James Square. Near a palace by the same name built by Henry VIII, St. James Square has anchored the seat of British power and prestige since the dawn of empire. It's a five minute walk to Regent Street, the Haymarket Theatre, the National Gallery or Trafalgar Square. A few minutes on is the other BP, better known as Buckingham Palace.

When the company pledged to create a $20 billion escrow fund to cover the environmental and economic damage of the spill, some in the Gulf region worried it might bankrupt BP. Some economists publicly suggested the company's assets should be frozen to ensure it could pay its debts.

No such thoughts rattled through the staid halls at No. 1 St. James, where investors were assured that the century-old company could weather the storm. BP's chief financial officer, Byron Grote, walked analysts through the numbers himself. BP would put up $5 billion by the end of this year, Grote explained, and $1.25 billion each quarter thereafter, until the full $20 billion was made available.

By eliminating dividends to its shareholders for the second half of the year, said Grote, the company would save the first $8 billion toward its goal. It would pick up $2 billion more by trimming its capital outlays over the same period to $18 billion from a planned $20 billion. With first quarter profits of $5.6 billion, the company was well placed to absorb the additional expenses going forward. Indeed, said Grote, BP would go ahead with plans to pay the Devon Energy Corporation, of Oklahoma City, $7 billion for some 840 billion gallons of oil in Azerbaijan and deepwater properties in the Gulf of Mexico and off the coast of Brazil. "As far as the Devon transaction goes," said Grote, "we would still expect to complete that in line with the agreement we reached with Devon earlier this year."

The second week of August, BP made its first payment—$3 billion—into the escrow account. "We intend to stand behind our commitment," BP's incoming chief executive officer, Bob Dudley, said in a prepared statement. By then, the company had spent $6.1 billion on spill containment and mop up operations, including wages for thirty-one thousand clean up workers and payments for more than five thousand "vessels of opportunity," mostly privately-operated fishing boats and recreational craft made available for clean-up. The total also included $320 million in damage claims paid out so forty thousand individuals and businesses.

Few in the region, though, seemed satisfied. In Orange Beach, Alabama, Mayor Tony Kennon said owners of restaurants, hotels and other local businesses all had the same complaint: their claims to BP were being scrutinized and squeezed so much that reimbursements weren't coming close to covering their losses.

"BP said they were going to make us whole," Kennon told *The Washington Post* in August. "Now they're finding ways to narrow down more and more who's getting paid." Kennon offered $500 out of his own pocket to anyone who felt BP had honored their claim in full. In fourteen meetings, he said, "Nobody has taken my up on that."

Many of the people BP employed for clean up operations worked long hours, braving blazing heat and sudden thunderstorms. They undoubtedly saved scores of miles of coastal areas, laying out and tending more than ten million feet of floating boom and skimming up thirty-five million gallons of oily water. They also conducted 411 so-called "controlled burns," torching more than eleven million gallons of crude oil at sea, the Coast Guard reported. That kept some oil off the beaches, while putting toxic chemicals and particulates into the air.

Kimberly Wolf showed up at the Buras meeting demanding to know what she and her neighbors were breathing. "We have a right to know what we're exposed to," the New Orleans community activist said, "so that we can figure out if we've actually killed the Gulf beyond repair or if repair is possible." She said she'd had enough of BP. "I want BP to get out of here," said Wolf. "All they need to do is leave a $20 billion check behind," she said to sustained applause. "We can handle it."

On the day before the meeting, the National Oceanic and Atmospheric Administration announced that seventy-four percent of the nearly two hundred million gallons of oil released from the Macondo well had either been skimmed up, burned off, evaporated or dissolved in the water. That left twenty-six percent unaccounted for in the ocean, estuaries, wetlands and shore. Combined with the oil thought to have been dissolved in the water, the Gulf had to somehow absorb 100 million gallons of oil. In an operation defined from the beginning, however, by confusion, misinformation and unanswered questions, there was skepticism in the Gulf over the precision of the estimate.

"I hope the report is right, but I don't put a lot of faith in it," Plaquemines Parish President Billy Nungesser told reporters. "I hate not to trust the government, but they haven't always been truthful through this whole thing." Shrimper Acy Cooper put it another way. "Somebody," he said, "is lying."

Overcoming that kind of distrust won't come easily, if at all. It certainly won't happen without a sustained, comprehensive and good faith effort to restore the health of the Gulf and make its residents and businesses whole. One key to that will be developing an accurate assessment of the damage the oil spill has done. The way that's done, by federal law, is through a Natural Resources Damage Assessment, or NRDA, a voluminous, legal catalog of the harm inflicted on the environment. The NRDA is part of a formal damage assessment and restoration process designed to help state, federal and tribal authorities deal with the aftermath of a catastrophic oil spill. Its being put together by the National Oceanic and Atmospheric Administration and the Department of the Interior, working closely with the natural resource offices from the five Gulf states—Texas, Louisiana, Mississippi, Alabama and Florida.

Thirteen working groups have been set up to pull together information on fish and shellfish, for example, birds and the broad range of affected habitat—from deep ocean waters and coral reefs to sea grasses, beaches, salt marsh and mud flats. Special attention will be paid to the twenty-eight species of marine mammals that make the Gulf their home. All of these are protected, under U.S. conservation laws, and six kinds of whales—sperm, sei, fin, blue, humpback and North Atlantic right whales—are listed as endangered. Similarly, at least four kinds of threatened or endangered turtles live in the Gulf—Kemp's ridley, green, leatherback and loggerhead.

The NRDA will also assess any lost human uses of the region's natural resources, such as recreational fishing, boating, hunting or visiting the beach. A complex and likely contentious process certain to take years, preparing the NRDA will involve extensive public input. Community meetings are planned throughout the fall so that regional residents can be heard. The integrity of the full process, though, is vital. In the minds of some, at least, it may already be threatened.

Weeks after the blowout, BP announced it would create a $500 million pool to fund scientific research of spill-related impacts on Gulf habitat and life, the very kind of information critical to the NRDA. By mid-June, the company had already awarded $25 million worth of grants to some of the region's most preeminent institutions, including Louisiana State University, the Florida Institute of Oceanography, affiliated with the University of South Florida, and the Northern Gulf Institute, a consortium led by Mississippi State University.

"It is vitally important that research start immediately into the oil and dispersant's impact, and that the findings are shared fully and openly," BP chief executive officer Tony Hayward said in a June 15, 2010 press release. To allocate the money, BP set up the Gulf of Mexico Research Initiative and named an international advisory panel to administer the grants. "We support the independence of these institutions and projects," said Hayward, "and hope that the funding will have a significant positive effect on scientists' understanding of the impact of the spill."

But regional advocates are concerned the investment could help BP and hinder sound science. For one thing, these advocates fear, it could provide the company with a deep bench of experts to testify on BP's behalf in the raft of lawsuits certain to come. And because such grants sometimes bar researchers from releasing their findings for several years, it could effectively sideline some of the most knowledgeable researchers in the region. It has the potential to compromise the quality and scope of the NRDA in ways that prevent a full and fair accounting of the damage the spill has done.

"Buying science is not science," said Mark Davis, director of the Institute on Water Resources Law and Policy at Tulane University Law School in New Orleans. "I don't see BP rushing around to fund actual research. I see them out there trying to buy the experts they're going to need to settle their case. And every expert they can buy is somebody who's not available to someone else."

The NRDA will become a crucial document. The conclusions it contains will provide the basis for a larger body of federal actions, legal decisions and restoration plans. The state and local offices that will help put it together, however, have little or no experience in doing so. Having witnessed the influence BP was able to exert over federal authorities in characterizing the spill to the public, Davis fears the states will be overwhelmed by BP's ability to influence the available research. "They've been controlling the flow of information since the day this began, and their liability is directly tied to the information base," said Davis, who spent fourteen years helping to shape programs and policies affecting Gulf waters and wetlands as director of the Coalition to Restore Coastal Louisiana. "If this case drags on in the courts for five years, and essentially nobody is doing responsible fisheries science because it's all going into litigation instead of our fisheries, then we're all screwed," he said. "We're all going to lack the capacity to make informed decisions."

It's essential that not happen. First, because the restoration stakes are so high. But also because we need accurate and detailed information—from independent scholars and researchers, as well as government agencies—to ensure that we learn the lessons from this disaster. Cynthia Dohner, southeast regional director of the U.S. Fish and Wildlife Service, says the spill has shown how little is understood about ecological processes fundamental to the Gulf and how much more needs to be learned to maintain a healthy and vibrant region. The NRDA and accompanying restoration work is an opportunity to develop that knowledge.

"We do not know, at this time, the extent of the injuries, but we believe that they will affect fish, wildlife and plant resources in the Gulf, and possibly in other areas across the country, for years and decades to come," Dohner, the Department of Interior's point person for the NRDA, said at a July 27 hearing of the Senate Water and Wildlife subcommittee. "The spill has illuminated the need for additional information about wildlife, fisheries and habitat, as we try to quantify the damage and understand the cumulative effects of the stressors that act on the Gulf Coast ecosystem."

What toxics did the oil and dispersant leave behind in the Gulf, for example, and in what quantities? Where did it make landfall? What impact has it had? What's happening in the deep water, on coral reefs and along the ocean and coastal bottoms where much of the oil is thought to have come to rest? What's the impact on marine and aquatic life, both juvenile and mature? What about spawning wildlife? What's moving up the food chain? What's happening in the marsh? What's left deep in the silt?

"Information on these impacts on economic activities, demographics, ecosystem services, as well as options for adaptation resilience planning are needed to help communities try to regain pre-spill productivity and social well-being," Marcia McNutt, director of the U.S. Geological Survey, said in June 15 testimony before the House Insular Affairs, Oceans and Wildlife Subcommittee. "Lessons learned from the *Exxon Valdez* oil spill suggest that a long-term—on the order of decades—multilevel ecosystem perspective will be essential."

As of late August, the National Oceanic and Atmospheric Administration was resisting making public much of the information it had collected in connection with the NRDA process. That included information it had received from, or shared with, BP. The public needs this information to have meaningful input into the process. NRDC has filed claims, under the Freedom of Information Act, seeking the release of this information. The request is pending.

The waterline doesn't lie. Eighteen feet above the floor, running just beneath the broad ceiling beams of laminated pine, it speaks in silent testament to the unyielding wrath of the deluge, the flood that crashed in uninvited to Ryan Lambert's fishing lodge, when Hurricane Katrina blasted through the levee on the other side of his backyard.

"Two-hundred-and-four-mile-per-hour winds. No electricity or water for eight and a half months. The water didn't recede for forty-three days." Five years on, his memory seared by the trauma, Lambert can reel off the raw statistics of the storm as clearly as he might describe his daily catch.

"There was no sound or color for six months," he said. "Everything was mud. Everything was grey. There wasn't a leaf on a tree. They were dead."

When a man watches everything he's built get taken by the sea, he either turns and walks away or rebuilds. Lambert rolled up his sleeves, threw his tools in the truck, and started hammering his dreams back together again. In six months he took off one day—to take his mom a birthday cake—until he'd rebuilt his lodge, nail by nail; until Cajun Fishing Adventures was open for business once more.

"I was possessed," he said. There seemed no better word for it.

Lambert sketched out the contours of his business late one night at the long wooden table where he serves hefty plates of fresh speckled trout, Gulf shrimp and hot biscuits to clients who come for a taste of the delta sporting life a world away from their homes in places like Little Rock, Charlotte and Nashville.

The business supports his family, Lambert explained, and twenty-two others. "Fourteen fishing guides, five duck guys, two do the cooking, one does the cleaning." Now he wonders how long he'll be able to make payroll. As reports of oil in the water have driven customers to scrub planned trips, Lambert has watched his business drop off like bait on a line tied to an eight-ounce sinker. "My business is down seventy to ninety percent from last year," he said, looking back on the country's deepest recession since World War II. "And that was my worst year ever."

Lambert is a tenacious optimist. Anyone who fishes to feed his family pretty much has to be. He can still find the fish; it's the customers that aren't biting. He's appealed to BP for damages, but that claim appears headed for court. After the long struggle to rebuild post-Katrina, heaving weathered what he'd hoped was the worst of the recession, Lambert sighed, shook his head and wondered out loud whether his can-do spirit will see him through once more. "I don't know if I've got another one of these in me," he confided. "I'm helpless. And it's a terrible feeling."

For Lambert and thousands of others like him, the Deepwater Horizon disaster is the third blow in an economic triple-whammy that has struck small businesses region-wide. For these entrepreneurs and their employees, still struggling to recover from Katrina's wrath and the economic downturn that's lingered since 2008, Gulf restoration means, first and foremost, paying the bills and keeping the doors open. "We're trying to survive day by day," said Plaquemines Parish Council Member Marla Cooper.

"We need to get this right," Senator David Vitter, a Louisiana Republican, said at the Water and Wildlife subcommittee hearing. "If we get it wrong, Gulf fishermen, recreational and commercial, and everybody else in my part of the world, could be seriously, economically crippled in terms of inadequate restoration of our wetlands and habitats."

Mabus will not wait for a full damage report. The region, he said, can't afford that. "We're going to restore the coast, both economically and environmentally," he pledged to the crowd in Buras. "It may be years before we know the full impact, but we can't wait . . . We need to get started."

As part of that effort, Mabus has enlisted the resources of every branch of the federal government that might have a role to play in helping to keep businesses like Lambert's viable. At the community meeting, tables lined both sides of the Buras auditorium with representatives from the Small Business Administration, the Internal Revenue Service, the Department of Agriculture, the Department of Housing and Urban Assistance, the Commerce Department and the Department of Labor, in addition to the Coast Guard and other agencies involved more directly with the oil spill.

The NRDA provides for emergency restoration measures that can begin while the overall damage assessment is underway. Mabus intends to make use of that provision. Nor does he pin his hopes on restoration to some distant project many won't live to see. The average U.S. Army Corps of Engineers project, he said, can take forty years from concept to completion. "We don't have forty years," he stressed. "We don't have forty months. Anything I come up with has got to be quick."

The task of restoration is urgent. It is critical that it be done in a way that is comprehensive, sound and forward leaning. By itself, though, that won't be enough. Until the country strengthens the safeguards we need to protect the Gulf and its people from harm, it will be only a matter of time before disaster strikes again. Whether through the steady environmental deterioration that has degraded this vital region for decades, or from the next Macondo-style blowout, further catastrophe awaits.

In order to reduce the chances of another blowout or similar disaster from occurring, to honor those who have died or suffered from this national catastrophe, and to restore one of the country's most vibrant and productive ecosystems, we need to rally around the Gulf and no forget about it. We need to institute needed change, not fall back into complacency. We must improve our oversight of the industry that has meant so much to this region, and to our country, for so long, yet has put so much at risk. And then we must rethink the grand bargain we all have struck, to feed our dependence on oil, by putting the quest for fuel above the health of the Gulf and the welfare of those who live here.

"A culture in which oil companies were able to get what they wanted, without sufficient oversight and regulation, that was a real problem," President Obama told reporters at a White House news conference in late May. "We have to make sure, if we are going forward with domestic oil production, that the federal agency charged with overseeing its safety and security is operating at the highest level. And I want people in there who are operating at the highest level, and aren't making excuses when things break down."

Obama took the vital first step of restructuring that agency, the now-defunct Minerals Management Service. An internal review has suggested additional reforms, many of which are well underway, aimed at improving on how offshore drilling rigs and production platforms are permitted, inspected and operated. The new agency, the Bureau of Ocean Energy Management, Regulation and Enforcement, is conducting an investigation of what went wrong, working with the Coast Guard as the joint Marine Board. Several committees in the Senate and House are conducting their own inquiries, as is the US. Department of Justice.

Obama has empaneled a special White House commission to investigate the Deepwater Horizon disaster and to recommend, by early January, needed changes to federal laws, regulations, agency structures, processes or industry practices. "We have an obligation to investigate what went wrong and to determine what reforms are needed so that we never have to experience a crisis like this again," Obama told reporters at the White House June 1. "If the laws on our books are insufficient to prevent such a spill, the laws must change. If oversight was inadequate to enforce these laws, oversight has to be reformed," he said. "Only then can we be assured that deepwater drilling can take place safely."

The National Commission on the BP Deepwater Horizon Oil Spill and Offshore Drilling is co-chaired by former Florida Governor Bob Graham, a Democrat who served three terms in the U.S. Senate, and William Reilly, who was Administrator of the Environmental Protection Agency under President George H.W. Bush. The president of the Natural Resources Defense Council, Frances Beinecke, serves on the commission with four other commissioners and the co-chairs. For the six-month period the commission is due to be active, Frances has walled herself off from NRDC's litigation and legislative activities and anything related to the oil spill.

Going forward, we must avoid the school of thought that has misguided so much of our policy over the past twenty years, a philosophy born of the notion that industry is the best keeper of its house, that trees are dirtier than SUVs and somehow collective governance is bad. Fortunately, we're beginning to hear some articulate leaders giving voice to a more credible view.

"A corporation, by its nature, by law, maximizes its economic self-interest. That's what they do. That's why they're formed. It's what they're for. It's what they're about. It's their very nature. It's their lawful obligation." These are the words of Senator Sheldon Whitehouse, a Democrat from Rhode Island. Standing in the U.S. Capitol in June, beside Gulf Coast residents who had lost family and friends on Deepwater Horizon, he laid out a clear case for the kinds of public protections we need.

"It's up to us in society to put the penalties in the right places, and in the right amounts, so that the decisions that they [corporations] make in pursuing their economic self-interest are the right ones," he said, "for workers, for families, for the environment and for society at large."

To begin, the country needs a set of safeguards to address the real risks posed by offshore drilling. The industry, and the hazards its work presents, have moved well beyond our current capacity for defending against those risks. While drilling in the Gulf expanded tenfold over the past two decades, the number of federal inspectors in the region grew by just thirteen percent, the House Oversight and Government Reform Committee found.

"The results: fewer than sixty inspectors are currently responsible for conducting over eighteen thousand inspections annually," said the committee's chairman, Congressman Edolphus Towns (D-N.Y.). "Over the last decade, MMS [Minerals Management Service] has essentially permitted the oil industry to police itself," he said at a June hearing on the issue. "BP and the other oil companies were essentially on the honor system . . . this is not an effective approach."

In late July of the Deepwater summer, the House narrowly approved legislation that would take vital first steps toward strengthening our oversight of offshore oil drilling and production. The bill, the Consolidated Land, Energy, and Aquatic Resources (CLEAR) Act, is an effort to enshrine lessons learned from nine hearings on the Macondo blowout before the House Energy and Commerce Committee.

The committee reviewed in detail a series of failures of the blowout preventer atop the well. If it becomes law, the CLEAR act would set minimum standards for such equipment. Blowout preventers would be required, for example, to have two sets of so-called blind shear rams—powerful blades capable of slicing through drill pipe and sealing it in the event of a blowout. While blowout preventers used in the Gulf commonly have twin sets of such rams, the blowout preventer used by Transocean on the Deepwater Horizon rig did not. It isn't yet known why the blowout preventer failed, but there had been reports of faulty communications and control gear in advance of the disaster. The CLEAR act mandates greater redundancy of such gear.

The bill also requires oil companies to take a number of steps aimed at ensuring that the well design and outer piping, or casing, are as reliable as possible. It requires, for example, at least three blowout barriers to be built into the design. BP's design had two. It requires the use of an adequate number of centralizers, parts attached to the outside of casing pipe to position it properly in the well hole before cement is poured in to seal the casing in place. BP used less than a third as many centralizers as its cement contractor recommended, a decision that may have been a factor in the blowout. The bill also requires that the integrity of the cement job itself be verified, a step BP skipped.

These procedures are already widely used in the offshore oil industry, but the legislation would require them by law. In addition, blowout preventers, well design, cement work and related procedures would have to be certified as safe by an independent inspector selected by federal regulators—at the oil companies' expense—instead of relying on the company's word, as has long been the case. The measure also makes oil company chief executives personally responsible for ensuring that well designs are safe. And it requires that blowout preventers have redundancy built in and that the company can promptly control and stop a runaway well should a blowout occur.

These measures are sensible, prudent and long overdue. Had they been in place a year ago, it's likely the Deepwater Horizon disaster would never have happened. It's hard to imagine how anyone could oppose these needed safeguards. And yet, the bill passed in the House by a margin of just 209-193, in part due to opposition to a provision that would create new conservation fees on oil and gas extracted from federal waters or lands.

I support such fees. As we've all learned or been reminded in recent months, offshore oil production puts irreplaceable habitat and wildlife at risk. Those risks aren't fully, or even nearly, accounted for in the price we pay for gasoline. The current Republican leaders in Congress, however, aren't on board. They call such fees an energy tax, which is something, even now, some in the party will not accept.

At least the bill passed in the House. Senate Majority Leader Harry Reid pulled companion legislation from that body just before senators headed home on their August recess, because it lacked the sixty votes needed to close debate and call a vote.

I wish more of our political leaders would take the time to go to the Gulf and see for themselves what this disaster has done. I wish they had flown, as I did, over the spill site and looked down to see rivers of black and brown crude oil flowing unchecked through those deep blue waters. I wish they would go there now and look Ryan Lambert in the eye and explain why it's acceptable to put this region at risk. I wish they would boat over areas that are shown as dry land on the charts. And I wish they would stand on Sal Gagliano's boat at the edge of a dying marsh and tell him and his children just exactly why we sacrificed a region, a way of life.

"This was a failure of policy long before it was an operational failure," said Tulane's Mark Davis. "The blowout itself really capped a long history of subordinating environmental, economic and cultural interests to the development of oil and gas. Changing that, from a policy standpoint, is absolutely essential if we're going to get anywhere. We have a shot at that, and that's all we have, is a shot."

As much as we need better safeguards, as much as we need stronger laws, we will never secure the long term health of the Gulf until we recognize it for what it is: a unique and irreplaceable treasure that's vital to us all, a source of daily sustenance, a sanctuary of life, a rich and resilient reservoir of the natural splendor that ties us to the larger world. It took two hundred million years to create these waters, this region and the diverse and reliant people who depend on it for their livelihood and their way of life. We have gambled its very survival on a callous and careless bet, staking its future on a belief rooted in the past, that we cannot, as a nation, rise above our addiction to oil. That, I believe, is wrong, and we have alternatives at our disposal that promise a better future for us and our environment.

Chapter 6

Beyond Petroleum

Two hundred million gallons is a lot of oil spewed out of the Macondo well in the three months it took BP to cap it. Most of that went into the ocean. If we somehow could have gathered that oil, and used it to fuel our country, how long would it have lasted us? The answer: six hours.

We consume eight hundred million gallons of oil—four times as much as the largest accidental oil spill in history—every single day. If we had a gas tank the size of the Empire State Building, we'd have to refill it every eight hours just to keep up.

In fact, if we could tap every barrel of oil known or thought to be lying beneath the entire U.S. portion of the Gulf of Mexico—all 1.7 trillion gallons of it—we could use up every drop, under current consumption rates, in just under six years. We could drill every square foot of the Gulf seabed, putting irreplaceable habitat and species at risk, stuff miles of steel pipe and cement down every well and bleed it dry. And we'd be out of oil before our next president leaves office.

That's not how it works, of course, because we get our oil

from many places. The arithmetic is simple enough, though, and the message couldn't be more clear: we're consuming oil at a reckless and unsustainable rate. We've got to break this addiction before it breaks us.

Far from some radical notion, that has actually been a strategic U.S. goal for four decades. President Richard Nixon called for "energy independence" and President Jimmy Carter summoned the nation to a national conservation and fuels replacement effort he said must be "the moral equivalent of war." President George W. Bush reiterated the objective in his 2006 State of the Union address. "Keeping America competitive requires affordable energy," Bush said. "And here, we have a serious problem. America is addicted to oil, which is often imported from unstable parts of the world."

Bush increased federal funding to help promote the development and use of hybrid cars, ethanol fuel from wood chips and other so-called biomass and new ways to harness the power of the wind and sun. "By applying the talent and technology of America," said Bush, "This country can dramatically improve our environment, move beyond a petroleum-based economy and make our dependence on Middle Eastern oil a thing of the past."

After decades of such talk, even at the highest level of American leadership, there's little to actually show for it. The Deepwater Horizon disaster, though, has given us a glimpse of the ugly future we're buying for ourselves with our persistent dependence on oil. The overarching lesson of the spill is this: we will take control of our energy future, at last, or it will take control of us.

"For decades, we have known the days of cheap and easily accessible oil were numbered," President Obama told the nation in June. "For decades, we have talked and talked about the need to end America's century-long addiction to fossil fuels. And for decades, we have failed to act with the sense of urgency that this challenge requires. Time and time again, the path forward has been blocked, not only by oil industry lobbyists, but also by a lack of political courage and candor."

These were no off-the-cuff remarks. It was the first time Obama had addressed the nation from the Oval Office. With oil gushing into the Gulf of Mexico, and no near-term way to stop it, the president wanted to use the symbolic seat of American power to make sure everyone understood the stakes.

"The consequences of our inaction are now in plain sight," he said. "As we look to the Gulf, we see an entire way of life threatened by a menacing cloud of black crude. We cannot consign our children to this future."

The good news is, we don't have to, if we make up our minds to do better. We didn't arrive at this point overnight, and it will take some time to change. By gathering ourselves as a nation, though, and making this a collective goal, we can cut our oil consumption from today's levels by eight percent over the coming decade. By 2050, we can cut it in half.

Doing that will make our economy stronger and our country more secure. It will reduce the carbon emissions that are warming our planet and threatening us all. And it will help to promote a healthy Gulf and create a brighter future for our children.

These projections are both conservative and sound. They rely on the use of technology that is either already available or rapidly emerging. They assume that our economy, and our population, will both continue to grow. And they stem from straightforward strategies and plans that will provide common sense options that enable all of us to do more of what we want to do with less. The key is to make our cars, homes and workplaces more efficient and to be smarter about how we design and build the places where we work and live.

As things stand now, we're on track to increase our oil consumption by 4.3 percent between now and 2020, according to a Natural Resources Defense Council analysis based on U.S. Department of Energy forecasts. By some lights, things could be worse. Those forecasts, after all, take into consideration improvements in auto and truck gas mileage, an increase in the use of fuels produced from biomass and reduced driving due to rising fuel costs.

If our goal, though, is to reduce oil consumption, not simply slow down the increase, we have to do better than that. And we can.

About seventy percent of the oil we consume is used for transportation, so the natural place to begin looking for savings is in our cars and trucks. We can cut our oil use by eight percent between now and 2020 by improving vehicle efficiency, creating alternatives to driving and promoting the use of cars and trucks that run on renewable fuels and electricity.

New cars in this country currently get 26.4 miles to the gallon, on average, when measured over a laboratory test cycle. That will increase to 34.1 miles per gallon by 2016, under the pollution and fuel economy policy Obama has already put in place, and thirty-five miles per gallon by 2020. Most of the improvements will come by using engines that are smaller but equipped with advanced fuel injection systems, coupled with more efficient transmissions than most cars use today. The result will be cars with plenty of pep delivered from engines that use less gas. That technology can be improved further if we promote more aggressive research and development. We can do a few things more, some of which are simple.

We can cut auto fuel consumption in this country by three percent tomorrow, if every car were equipped with good tires, properly inflated, and fuel efficient motor oil that lubricates engines better than conventional oil. We can also build cars that are strong yet lighter, using advanced materials and techniques. And we can design cars with cleaner aerodynamics, to reduce drag.

Based on analysis from the Massachusetts Institute of Technology, the NRDC calculates that our gasoline-fueled cars, pickups, minivans and SUVs, can deliver fifty-five miles per gallon, on average, by sometime around 2025. That's an improvement of 28.6 miles per gallon over today's performance. With nearly 140 million passenger cars in the country, that represents a substantial savings of oil, and we can get similar gains from delivery trucks, tractor-trailers and even heavy off-road equipment used for construction and other purposes.

Taken together, these motor vehicle efficiency improvements can reduce our total oil consumption by about fifty million gallons a day—about six percent less that what we use today—by 2020. That's the equivalent, by the way, of the output of roughly twenty five Macondo-sized wells. We can get tremendous savings simply by scaling up the use of existing hybrid electric cars and other fuel-efficient technologies. A hybrid has a conventional gasoline engine, as well as an electric motor. The car's computer controls the drive system, using the electric motor when practical, such as when driving in traffic or cruising at highway speeds when little power is required to maintain momentum. The gasoline engine switches on only when full power is required.

The hybrid saves energy in two ways. First, electric cars are vastly more efficient than those run by internal combustion engines. That isn't environmental policy: it's physics. The typical gasoline engine throws away about two-thirds of its energy through friction and heat. Lots of metal parts rubbing against each other at high speeds; lots of hot gas blowing out the exhaust pipe. Gasoline powered cars lose another seventeen percent, on average, with the engine idling while the car's not going anywhere. When we put ten gallons in the tank, in other words, only about two are actually being used to move the car down the road. The rest is wasted by the equipment; it's largely physics, and it can't be helped.

Electric motors, by contrast, apply seventy-five percent of the energy they consume to actually moving the car. They are inherently more efficient than internal combustion engines—they contain fewer moving parts—and they don't idle when the car's not moving. The other way hybrids save energy is by capturing it from the car itself. Generators recharge the battery when the car is running on gasoline and when the car has sufficient inertia— going downhill, for example, or coasting—to operate the generator without slowing the car. Generators also kick in when brakes are applied, harnessing some of the energy normally wasted as heat. That's why one hybrid, the Toyota Prius, gets fifty-one miles to the gallon in city driving, or forty-eight on the highway, according to the latest EPA estimates. Right now, hybrids make up only about one in fifty of the passenger cars on the road. Their ranks, though, are rapidly growing.

The Department of Energy lists thirty-four hybrid cars available in the United States. The Toyota Prius is the stingiest. In second place: the Ford Fusion, which gets forty-one miles per gallon in the city or thirty-six on the highway. The Honda Insight gets forty miles to the gallon in the city and forty-three on the open road. Chevrolet even makes a hybrid pickup truck. Its full-sized Silverado gets twnety-one miles per gallon in the city and twenty-two on the highway.

As hybrids continue to evolve, they will take advantage of many of the same drivetrain and platform improvements benefiting conventional gasoline-powered cars, as well as efficiency gains in hybrid components. We estimate hybrids can account for about one in every five cars on the road by 2020.

Toyota last year became the first manufacturer to offer a plug-in hybrid, which can be fully charged from a home or garage. General Motors plans to introduce its plug-in hybrid, the Volt, in 2010. The Prius plug-in gets thirteen miles on a charge. At that point, the gasoline engine must kick in. It's a start, though, toward creating a fully-electric option for local errands and short commutes. The NRDC estimates that plug-in hybrids and fully electric cars can make up ten percent of all new passenger cars by 2020, twenty-five percent of the new car fleet by 2030 and replace the sales of new gasoline-only cars completely by 2050. At that point, the NRDC expects, our cars will be getting the gasoline energy equivalent of about one hundred miles per gallon.

As we move to electrify our cars, we can also move to electrify our trains. Around New York, for example, most the commuter trains have been switched to electricity over the years. These electric trains are faster, quieter and vastly more fuel-efficient than the diesel-burning equipment they replaced. Some of the trains that run longer routes still use diesel. While adding the electric third rail for those routes will take time, it can and should be a part of our transportation future.

Public policies that help to promote the development of the next generation of hybrid cars could benefit our economy in another way, by building a foundation for revitalizing the

American automobile industry. One of the reasons we burn so much gasoline in this country is because so many of our cars are gas hogs. In 1991, the on-road fleet of American passenger cars averaged 21.1 miles per gallon. By 2008, the average car was getting 22.6 miles per gallon in real world driving. That's an improvement of 0.9 percent. When it comes to driving more efficient cars, we lost twenty years.

Americans who want to drive more efficient vehicles have looked largely to Japanese and European automakers. That's one reason for our global trade deficit. In the five years between 2005 and 2009, Americans imported $600 billion more worth of cars and parts than we shipped out of the country, or about eighteen percent of our total trade deficit—$3.4 trillion. During the same time, oil imports accounted for $1 trillion of our trade imbalance: that means that, over the past five years, oil and auto imports have accounted for $1.6 trillion of the U.S. trade deficit, or forty-eight percent of the total.

Nor is this a recent development. Between 1994 and 2009, cars and oil combined to account for $4.7 trillion of the nation's trade deficit, about two-thirds of the entire trade gap. That's a drain on the American economy we simply can't afford. If the United States would get serious about manufacturing a world-beating fleet of fuel-efficient cars, the country could begin taking a bite out of that trade deficit from both ends. The result would be better earnings for millions of Americans and trillions of dollars that stayed in this country. What on earth are we waiting for?

This isn't George Jetson stuff. It doesn't count on some revolutionary technology we can't yet foresee. This is what we can accomplish together if we invest wisely to improve on technology that's here today and shift to more efficient vehicles than those now on the road.

Freight trucks and other heavy equipment can attain comparable savings through efficiency gains. The average new heavy duty truck gets about six miles to the gallon. The NRDC estimates that could improve by sixty percent by 2020. A recent study by the National Academies of Science concludes that technologies available by then could as much as double the fuel efficiency of heavy trucks. The NRDC conservatively estimates that that entire fleet of freight trucks can average twelve miles to the gallon by 2050.

These efficiency gains can, by themselves, reduce our oil consumption dramatically. Over time, we can make further cuts by developing practical options for moving goods from place to place, getting to work, doing our shopping and visiting our friends. Public transportation—buses, streetcars, subways, commuter rail lines and other sources—save the United States the equivalent of 1.4 billion gallons of gasoline every year, according to a 2007 study by ICF International, a consulting firm based in Fairfax, Virginia. In addition to being more efficient at moving people around than personal automobiles, commuter rail systems can run on electricity and buses can run on natural gas.

Every day, millions of passengers demonstrate a preference for mass transit over driving when they have that option. Washington, Atlanta, Chicago, Seattle, New York and other cities have invested in dependable, efficient public transit systems. For millions of Americans elsewhere, those kinds of transit options are too limited to be of value. We haven't invested enough to provide clean, safe and dependable public transportation options for them. By doing so, we could dramatically reduce our oil consumption.

In some ways, we've actually gone backward on this score. Even Baby Boomers from American cities ranging from New Orleans to Boston can remember electric street cars and trolleys that tied suburbs to the city and linked it all to the railroad. And yet, many of those practical, convenient, economical and efficient transit systems were torn up in the 1950s and 1960s, as we invested in highways, roads and cars. Unfortunately, highways now get eighty cents of every federal dollar spent on transportation in this country. Much of that goes to build new roads in sparsely populated areas. Yet eighty percent of the economic activity in this country occurs in the nation's hundred largest cities. If we devoted a proportionate amount of our transportation dollars toward building and improving public transportation there, we'd have more options for supporting the kind of clean, energy-efficient economic growth that can genuinely improve our standard of living.

In the freight sector, tremendous amounts of fuel can be saved by shifting even a modest portion of goods now shipped by truck to rail. Mile for mile, shipping a ton of freight by rail takes just fifteen percent as much energy as shipping it by truck, according to a 2007 Department of Transportation report entitled *Best Practices Guidebook for Greenhouse Gas Reductions in Freight Transportation*. Rail isn't always the more practical option, depending on the commodity, the distance being traveled and the proximity of rail routes to the ultimate destination. But if just ten percent of the goods now shipped by truck were to travel by rail, we could save about seven million gallons of diesel fuel a day.

And yet, the trend is moving in the opposite direction. Truck volume is on track to increase by sixty percent over the next twenty years, while rail use is expected to grow by just thirty-nine percent, according to Department of Energy projections. The gap is being widened by federal subsidies—eighty percent of all federal transportation money goes to our highways.

In some cases, we can cut oil use simply by coordinating

transportation and lifestyle options when we plan our cities and communities. Employing what some call "smart growth" strategies, planners are siting common destinations—schools, shopping centers, offices, theaters, apartment buildings and the like—with the routing of roads or the availability of mass transit, parking facilities, pedestrian walkways or bike paths. The idea is to build communities where people don't have to get in a car every time they want to catch a movie, grab a bite or get to work.

In communities planned with these goals in mind, residents and visitors make fewer car trips, and the car trips they do make are shorter than the national average, meaning less fuel is used even as people stay mobile, according to a 2008 report, "Growing Cooler" by the Urban Land Institute, a Washington-based research and analysis outfit.

Atlantic Station, for example, is a 138-acre development near downtown Atlanta. It was planned around a compact, transit-oriented design with an emphasis on public transportation and pedestrian access. The ten thousand people who live there drive up to sixty percent less than the average Atlanta resident, according to a report prepared for the EPA.

Private employers have begun offering to pay employees to forego the company parking spot. That reduces construction costs on new offices and factories. Employees pocket so-called "parking cash-out" payments, then find alternatives: car pools, mass transit or bicycling, for example.

By promoting the kinds of efficiency gains and transportation alternatives discussed here, we could cut our country's overall oil consumption by eight percent by 2020, by thirty percent by 2030 and by about half by 2050, the NRDC estimates. These are, I should stress, conservative estimates. They don't take into account any breakthrough technologies that might emerge over the next four decades. When we look back at the technologies that didn't exist just thirty years ago—cell phones, the Internet, laptop computers, GPS or the ability to drill for oil beneath ocean waters two miles deep—yet are now taken for granted as part of our daily lives, we know that surely fuel-saving technologies can emerge in the years to come that we can scarcely conceive of now. Until then, there's more that we can do.

We can replace some of the oil we do burn with what is called cellulosic fuel. That is a kind of fuel produced from farm waste—corn cobs, wood chips or wheat straw, for instance— municipal waste, such as household garbage and paper products. It can also be produced from agricultural products, such as switchgrass and fast-growing trees such as poplar and willow, grown specifically for fuel on degraded land that cannot grow food crops. Yet another source for such fuel is residual vegetation—corn stalks, for example—left over after harvest. Some vegetation, of course, needs to be left on fields to replenish them, but much of what is now commonly removed and burned as waste could be a source of fuel.

The NRDC estimates we could produce one hundred million gallons a day of high quality biofuels from these sources by 2030. This kind of fuel generally burns with roughly two-thirds the fuel content of gasoline, so the replacement isn't gallon for gallon. Biofuels, though, could replace more than sixty million gallons a day of oil in this country within twenty years. That's nearly eight percent as much as we currently consume.

One type of biofuel is ethanol. Right now, most American ethanol is made from corn. The NRDC does not regard corn ethanol as a sustainable alternative to oil, chiefly because corn production itself uses such large amounts of fuel, land, water and agricultural chemicals. In fact, fertilizer runoff from heartland cornfields contributes to nitrogen loading in the Mississippi River, which feeds the algae blooms that deplete the oxygen creating dead zones in the Gulf. The country, however, is on track to expand its corn ethanol production to about forty million gallons a day, through existing facilities or those scheduled to come on-line in the near future. That is separate from the cellulosic fuel production endorsed by the NRDC.

Diesel fuels can also be made from a variety of renewable sources. Waste cooking oil from restaurants and food processing plants, for instance, is collected in some cities and converted to bio-diesel fuel, rather than thrown away. Some perennial crops that don't require yearly tilling can be used to make diesel fuel. Dozens of laboratories around the country are exploring ways to create diesel fuel from microscopic algae. Technologies that for many decades have converted coal into liquid fuels can also effectively turn almost any form of biological waste, even sewage, into a wide range of fuels.

There are many options and we're only beginning to learn which ones will be the most successful. But the history of our country is that, with the right incentives, American business can deliver astounding products. That's why several states are already creating incentives for low-carbon fuels that will use less petroleum, an effort the NRDC is supporting and urging the federal government to follow.

"American entrepreneurs and workers have the ingenuity and grit necessary to break this addiction," Senator Jeff Merkley, a Democrat from Oregon, said in July when he and three other senators introduced the Oil Independence for a Stronger America Act. The bill calls for reducing U.S. oil consumption by 250 million gallons a day by 2030, by encouraging many of the measures discussed in this chapter. "The challenge is whether politicians in Washington are willing to choose American strength over vulnerability."

Change is seldom easy. When it comes to reducing our reliance on oil, though, at least some solutions are readily defined. And they are well within our grasp. As we set out to find ways to reduce our reliance on oil, in fact, we will accelerate our progress in ways we can't now foresee. That's how innovation and human nature works. The more we focus on solving a problem, the better we become at doing so. Examples from our own history abound.

In the mid-1970s, the state of California took note of a wide variation in the amount of electricity consumed by home refrigerators. Sacramento begin to implement efficiency standards. In 1992, the EPA got into the game, offering a $30 million reward to the manufacturer that could meet a specific standard. Whirlpool took the prize and by 2006, the nation's refrigerators were using seventy-five percent less energy than they were using twnety-five years before.

That story and others are told in the 2010 book *Invisible Energy* by my NRDC colleague, David Goldstein. More than a brilliant overview of what's possible through efficiency gains, the book is a tribute to the power of American innovation. American ingenuity can break our addiction to oil and begin moving this country to safer, cleaner, more sustainable sources of power and fuel. Goldstein leaves no doubt about that.

Over time, he writes, sustained, incremental reductions can cut our reliance on petroleum even more dramatically than we now realize. If fuel economy, by itself, were to improve steadily by 3.5 percent a year, we would see a seventy-five percent reduction in use by 2050 from the new passenger car fleet alone.

Imagine what that might mean. We could be out of the business completely of relying on foreign countries for strategic fuel. We could be investing in this country hundreds of billions of dollars each year that are now going abroad. We could reduce, dramatically, the risks we now impose on special places like the Gulf of Mexico. And we could do it in a way that creates millions of American jobs producing the clean energy technologies of tomorrow, opening the door to decades of success in fast-growing markets worldwide.

On May 25, 1961, President John F. Kennedy stood before a special session of Congress and summoned the nation to commit to putting a man on the moon before the decade was out. Eight years later, that mission accomplished, Americans embraced the national pride and the enduring benefits that came from harnessing the country's collective resources to the attainment of a common goal.

The nation that won World War II and the Cold War, that developed the Internet, the personal computer and put a man on the moon, has what it takes to break its dangerous and destructive addiction to oil. We must make up our minds to do so.

Epilogue

For most of his life, David Arnesen has spent summer nights on the Gulf of Mexico, fishing for the mackerel, snapper, grouper and other seafood he sells to support his family. This year, though, was different. With oil in the water, fishing grounds were closed. And so, on a hot night in August, he found himself in the Louisiana hamlet of Buras, in a stifling auditorium packed with fellow citizens and federal officials, awaiting his turn at the microphone.

"I know right now oil is the most valuable thing on the planet," he told Navy Secretary Ray Mabus, who was putting together a plan to help the Gulf and its people recover from the Deepwater Horizon oil disaster. "Maybe one day food will be."

By then, Arnesen was living by himself. His wife Kindra had packed up their children and moved them near New Orleans, out of fear that airborne toxins from the BP blowout were causing sickness and rash. Arnesen stayed back to find a way to pay the bills any way he could.

"We may have millions of pounds of food that we will never be able to put on the plate anymore," said Arnesen. "How many millions of pounds of food can we afford to give up before there's not going to be enough?" He glanced toward Kindra, who'd driven down to attend the meeting, then turned back to conclude. "We can't afford to give up food sources. It's too big a trade-off."

Oil is valuable, precious even, because of all we rely on it to do. For far too long, though, we've given up too much in our thirst for more. Too much of our national wealth, our natural environment, our security, our safety, our health. It's become too big a trade-off.

That's the lesson we must take from the BP oil disaster. That's the plea from the beating heart of the Gulf. We must rethink our dead-end dependence on oil and the industry that provides it to us. We must take better care of each other. We must take better care of our planet. We must break our addiction to this dangerous and destructive fuel once and for all and begin moving toward cleaner, safer more sustainable sources of energy.

Our most urgent responsibility is to Arnesen and millions of Gulf Coast residents like him, whose lives have been thrown into turmoil because decades of poor stewardship of this vital region have been capped off by the worst peacetime oil spill in history. We owe the Gulf, its wildlife and its people, first and foremost, a comprehensive and honest reckoning of the damage done by this catastrophic spill. That is the work of government officials—federal, state and local—as well as scholars, scientists, environmental justice advocates and others who have the authority, and the obligation, to ensure that the Natural Resource Damage Assessment tells it like it is. That is the only way we can begin the long process of restoring trust among people whose faith in our institutions was shattered by the response to Hurricane Katrina and has been tested once more by the confusion, unanswered questions and misinformation that added insult to injury when the well blew out.

Fairly cataloging the damage, though, is only the start. We must, as a nation, stand by this productive yet wounded region and its people long enough for economic and environmental recovery to take hold. It's BP's job to write the checks to make that happen. It is the role of the federal government to make sure it is done right.

"Our job is not finished, and we are not going anywhere until it is," President Obama pledged in Panama City, Florida on August 14. "My job is to make sure that we live up to this responsibility, that we keep up our efforts until the environment is clean, polluters are held accountable, businesses and communities are made whole, and the people of the Gulf Coast are back on their feet."

For that to happen, the Mabus restoration plan is going to have to address the long-term issues of the delta. The vanishing wetlands. The harm done over the decades by cutting canals and pipeline routes through the marsh. The redirection of river sediment by levees. The nitrogen-loading of the Mississippi River from farm and municipal runoff hundreds of miles upstream. The BP blowout did not cause this damage. But it has highlighted the fragile state of this irreplaceable delta habitat upon which the health of the entire Gulf depends. We have driven this essential delta to the brink of disaster. We need, as Mabus said, a healthy Gulf. That will require a resilient and strong delta. By late August, hopeful signs were emerging. Young thin shoots of spartina grass, roseau cane and black mangroves began poking up from oiled wetlands, an indication that the marsh might be regenerating itself in places, a natural healing that needs our help. "I know we're going to come back from this," said Mabus. "The Gulf is not going to be forgotten."

To the extent oil companies continue to mine the Gulf for oil and gas, we must insist that they do so safely. It is the role of government, acting on our behalf, to put safeguards in place to protect our environment, our health and the welfare of the workers who produce oil for us. For decades we have allowed our politicians to diminish and even denigrate the role of these needed protections and the people who enforce them. It was a shortsighted and ultimately pernicious approach, and the Macondo blowout is part of the price we are now paying.

"Some of our watchdogs weren't watching close enough," said Arnesen. "We need someone to push harder to make sure it's done safely."

Fortunately, we have a functioning democracy that can learn from its mistakes and correct its course. What the Deepwater Horizon debacle has taught us is that our free-market system, for all its virtue and strength, needs effective public oversight. It is an investment in our future economic well-being. Just as we must invest in communications, research and development and transportation, we must also invest in responsible public oversight, setting the relevant rules of the road, building agencies with the authority to enforce those rules and empowering people with the knowledge, the resources and the tools they need to defend the public good. We will rise to that responsibility or we will continue to put ourselves and our world at growing risk. And we need more disclosure so that communities can participate in oversight. These fossil fuel resources belong to the public—that's why the oil companies pay royalties—and the impacts of extraction fall on the public. The public deserves to know what is going on and have access to the information. The days of everything important happening behind closed doors should be over.

Finally, we must recognize the role each of us plays in this disaster. The oil companies in general, and BP in particular, have much to atone for, no doubt. It is the American public's demand for oil, however—eight hundred million gallons of it every single day—that has driven the industry into deeper and more perilous Gulf waters in search of new sources of crude. And it has been our willingness to go anywhere, bear any risk and pay any price for that oil that has funded this increasingly dangerous enterprise. We must make drilling and producing offshore oil safer right now. In the long run, we must make it obsolete. We must reduce our dependence on oil. We know how to do this. We have the technologies to make our cars, trucks and trains use two or three times less oil. And, after decades of delay, we have finally started down that path. Clean energy systems that rely on wind, water, the sun or the Earth itself are well-established. We know that homes in communities that offer alternatives to hopping into a car to do almost anything are popular, sell well and hold their value better than homes in communities with fewer transportation options. All of this we know: it is just a question of our will to make it happen.

And what we do not yet know, we can discover. We can endeavor to find new technologies and systems, scarcely considered today, to change how we live and work, to traffic in ideas without moving goods and people around or to move in different ways, to find energy sources that do not put public health and natural resources at risk, to store and transmit energy.

Eight hundred million gallons of oil. One fourth of every drop produced everywhere in the world. Enough to fill the

Empire State Building nearly three times. That's the problem. That's the role each of us played in sowing the seeds for the Macondo blowout.

"That's part of the reason oil companies are drilling a mile beneath the surface of the ocean," Obama told the nation in his address from the Oval Office June 15. We are losing jobs to other nations that are pushing forward to create the clean-energy technologies of tomorrow. We are sending nearly a billion dollars a day overseas to buy oil. And we are looking at the Gulf of Mexico and its people and seeing a region and the way of life it supports "threatened by a menacing cloud of black crude," he said. "We cannot consign our children to this future."

A hundred ten years ago, American ingenuity, entrepreneurship and resolve combined to change the world at Spindletop. A gusher was tapped, an industry was born, our country, and the world, would never be the same. I'm not sure the folks who drilled Spindletop were Mozart fans, but let's imagine, for a moment, they were. To hear *The Magic Flute*, they would have needed tickets to a concert hall. Today, we can put every classical opera and symphony ever written on an iPod smaller than a deck of cards, and still have room for a respectable history of jazz, blues and rock 'n' roll. If they'd wanted to send a letter to London in March, it would have done well to get there by June. Today the same missive takes only the time required to tap it out as an e-mail. And if they wanted to tell someone in Topeka the big news from Spindletop, the local telegraph office would have been their best bet. Today they'd punch in a few digits on a cell phone.

The oil industry has come a long way too. At the same time, we still basically drill into the ground to get oil, and then refine it into fuel, lubricants and petrochemical feedstocks, much as we did a century ago. For all the changes in the automobile industry, the internal combustion engine still grinds out horsepower pretty much as it always has, wasting six out of every ten gallons of gas that it burns.

Our history shows we can do better than that. And the Macondo blowout shows that we must.

Acknowledgments

Thank you to the people from the Gulf who joined with NRDC since the BP gusher exploded. LaTosha Brown and Bryan Parras of the Gulf Coast Fund provided guidance and joined forces with NRDC to open the Gulf Resources Center in Buras, Louisiana. Diana Huhn of Bayou Grace Community Services and Rosina Philippe of Grand Bayou Community United have supported our work at each turn. Ryan Lambert took me out on the water. And Kindra Arnesen, the wife of a fisherman from Venice, Louisiana, infused fury in the fight to protect the health of Gulf families.

People like Kindra were only just beginning to recover from the devastation of Hurricane Katrina when the oil spill hit their shores. Their resolve in the face of this disaster astounds me.

Thank you to John Adams, founder of NRDC, and NRDC President Frances Beinecke, both visionary leaders who tackle the toughest environmental fights and believe better solutions are possible. So many of our colleagues at NRDC contributed to this book. Our staff works every day to secure a healthier Earth, one where people, animals, plants, and the systems we depend upon might survive and thrive.

The NRDC oceans team has led the response to the BP

blowout. They have brought decades of expertise and passionate advocacy to bear in the midst of this crisis. Many thanks to Sarah Chasis, Lisa Speer, Lisa Suatoni, Regan Nelson, and Alison Chase.

NRDC's public health and urban teams have also been on the ground in the Gulf, as they were after Hurricane Katrina, to investigate health concerns that others ignored. Thanks to Gina Solomon, Sarah Janssen, Miriam Rotkin-Ellman and Al Huang.

NRDC's oil and transportation teams have been leading the charge to reduce our dependence on oil. Many years of effective advocacy contributed to the progress we are finally seeing to make our cars, trucks and transportation systems more efficient and cleaner, and to laying the ground work for further advances. I am grateful to Ashok Gupta, Roland Hwang, Luke Tonachel, Deron Lovaas, Colin Peppard, David Doniger, Kaid Benfield, and Nathanael Greene.

NRDC's communications team has worked tirelessly to make sure the stories, struggles, and concerns of Gulf residents are heard in Washington, D.C. and around the nation. Special thanks to Rocky Kistner and Anthony Clark for the successful launch of the Gulf Resource Center.

Several other NRDC experts have shaped our response to the Gulf crisis and informed this book, especially Wesley Warren, David Pettit, and David Goldston.

Both NRDC and the NRDC Action Fund are fortunate to have dynamic and engaged trustees, who form a tight partnership with the staff and enhance our work. And NRDCs dedicated members, who remain a driving force behind our greatest victories, keep us strong as we fight those who would continue to pollute.

John Oakes, our publisher, brought the idea for this book to Doug Barasch, NRDC's *OnEarth* editor, and Phil Gutis, NRDC's communications director. Jennifer Powers provided vital insights into the book publishing world and our lead researchers, Matthew Eisenson, Judi Hasson and Emily Deans, worked tirelessly to track down key figures and facts.

Bob Deans' firsthand experience shaped the book, and his journalistic background enriched the text.

Finally, thanks to my wife, Fritz, for her vision and friendship.

Resource Guide

To help strengthen needed safeguards, restore the Gulf and break our addiction to oil, follow Peter Lehner's blog for the latest developments at: http://switchboard.nrdc.org/blogs/plehner/

For additional information about the Gulf and the Deepwater Horizon disaster:

Oil: Money, Politics, and Power in the 21st Century, by Tom Bower (Grand Central Publishing: New York, 2010).

The Tyranny of Oil: The World's Most Powerful Industry: And What We Must Do to Stop It, by Antonia Juhasz (William Morrow: New York, 2008).

The Prize: The Epic Quest for Oil, Money and Power, by Daniel Yergin (Free Press: New York, 2008).

The Natural Resources Defense Council
www.nrdc.org

The official site of the Deepwater Horizon Unified Command response team
deepwaterhorizonresponse.com

The National Oceanic and Atmospheric Administration's Deepwater Horizon response site
http://response.restoration.noaa.gov/deepwaterhorizon

U.S. Fish & Wildlife Service Deepwater Horizon response site
http://www.fws.gov/home/dhoilspill/index.html

Louisiana State University School of the Coast and Environment oil spill page
http://www.sce.lsu.edu/oilspill.shtm